Stone House Construction

Dedication

In memory of Doug Swing
who opened up a realm of possibility
where ideas could be acted on, emerging as another reality.
Doug, your work continues to inspire,
and I miss you.

Stone House Construction

SARAH GUNN

CSIRO
PUBLISHING

National Library of Australia Cataloguing-in-Publication entry

Gunn, Sarah.

Stone house construction/by Sarah Gunn.

9780643096370 (pbk.)
9780643106857 (epdf)
9780643106864 (epub)

Includes bibliographical references and index.

Stone buildings - Australia.
Building, Stone - Australia.
Stone industry and trade - Australia.

691.20994

Published by
CSIRO Publishing
36 Gardiner Road, Clayton VIC 3168
Private Bag 10, Clayton South VIC 3169
Australia

Telephone: [+613] 9545 8555
Local call: 1300 788 000 (Australia only)
Fax: +61 3 9662 7555
Email: csiropublishing@csiro.au
Web site: www.publishing.csiro.au

Front cover: Sandstone contrasting with the local stone around a window, and a plank as a simple means of working above it. Window stonework by the author and her husband, continuing a family project.

Back cover: Rubble stone steps.

Set in Adobe Minion Pro 11/13.5 and Adobe Helvetica Neue LT
Edited by Anne Findlay, Editing Works Pty Ltd
Cover design by Alicia Freil, Tango Media
Text design by James Kelly
Typeset by Desktop Concepts Pty Ltd, Melbourne
Printed by Ingram Lightning Source

Feb26_RP_ILS

Contents

Preface

When the challenge to design a stone house came my way, I went looking for a reference book dealing with stone in our region to help me with the project. I did find a number of books, each one helpful in its own way. However, I still had in my mind's eye the book that would better describe the building process we were about to launch ourselves into.

Stonework is a complex and detailed area of expertise, yet we took it on without the benefit of trade experience, as do many people. Somehow we were able to achieve impressive results, doing things we had never done before.

This is the book I was looking for then, the book I was imagining, or as close as I can get to it. The photographs are of places and people we found on our path to extend our knowledge. This journey includes learning from the work of many architects and builders. Those who are known are credited at the end of the book.

Perhaps there is no real short cut to learning by experience. The closest thing is learning from the experience of others. On returning to photograph stone buildings that I had admired before, I reached another level of admiration when I understood better the subtleties of stone construction. My hope is that readers will develop a similar level of understanding and inspiration from this book.

Sarah Gunn, 30 May 2011

Acknowledgements

To all those people who so generously allowed us to photograph their houses, gave us directions and shared with us their stories, I wish to express appreciation. I offer apologies to all whose names are not included. I can at least mention Patricia Brown in the Bairnsdale area, and Dan Clarke in Narrandera, for historical information about their local buildings.

I am grateful to Professor John Scrivener for helping make this study an academic pursuit. Also to Ted Howard, Lynne Hume and David Miller, for helping me to begin writing a book.

As for the house near Hepburn Springs, I owe thanks to my mother for getting us all started. As a design reference, I used a photo of a castle my mother liked (Olofsborg) in a book (Tuulse 1958). I wish to acknowledge the many friends who came to working bees for our building project, or helped on weekends, as well as those who gave moral support, especially Denison Deasey, Paul Evans and Pen Alexander. I also offer thanks to my sister Anne for her patience in proofreading my first drafts of the book, and for her many helpful suggestions. Thanks to John Endacott for describing working with granite, to Bert Behnke for discussing tools in general, and to Graham James for showing up out of the blue (with Maggie Fooke) and offering to proofread. Also to Janis Brodie for typing, to Nerida Osborne for computer assistance with photos, to Trevor Scott for proofreading a final draft, and many thanks to Ted Hamilton for all his help getting my material ready for publishing.

More than anyone else my late husband, Doug, helped to make the first idea of a book grow into what it has become. My journey into stone construction was his journey too. He used his knowledge of hard rock mining and applied it to quarrying the local stone. Without him, I could not have found all the stone houses photographed in this book. I am so thankful for the dialogue I had with him, which helped me to work out many of the challenges of building.

Figure 1.1: Camelot, a flat-roofed, circular central space defines this house, with a cantilevered stair to roof, Castlecrag, New South Wales.

'The floor was rough and stony; the walls full of projecting corners;
the roof in one place twenty feet high, in another endangering his forehead;
while on one side a stream, no thicker than a needle ... flowed down the face of the rock.'

George MacDonald (1872), *The Princess and the Goblin*

1

Stone house design

A well-built stone building withstands the test of time. A humble stone cottage from a bygone era can engage our interest. A grand old stone building can excite our imagination much more.

Our home is our castle. We dream up such a castle when designing a house. Will it be grand, or humble? Will it attract attention or be private? Above all, it should be comfortable to live in.

Thermal comfort

People may say stone houses are cold. But they don't have to be. Although not a good insulator, stone has an excellent capacity for heat storage. A stone house with a good aspect catches sunlight in winter but not in summer, so is comfortable to live in year round. The principles of house design described below (Table 1.1 and Figure 1.2) take advantage of this capacity to store heat, known as thermal mass. A measure of thermal mass is the response period of an element (the time it takes to absorb and re-radiate heat, based on a pulse of heat moving through it). The response period is very long for stone walls, with some variation for different kinds of stone, and it increases with wall thickness. Mudbricks and fired bricks are also thermal mass materials.

Letting sunlight into the house for warmth (but not too much) and getting the right balance of light and shade are the qualities that make a good living space. A benefit of good design is that the house should maintain a comfortable year-round indoor temperature without resorting to costly heating and cooling methods.

Table 1.1: Heat transfer: comparison of wall types

WALL CONSTRUCTION	THERMAL TRANSMITTANCE U (w/m².deg.K)*	RESPONSE PERIOD P (hours)*
Timber/air (100 mm)/ plasterboard	1.82	4
Timber/air (50 mm)/mineral wool (5 mm)/plasterboard	0.5	7
Brick (veneer)/air (140 mm)/plasterboard	1.98	16
Brick/cavity (40 mm)/brick	1.88	31
Mudbrick (300 mm)	2.47	43
Granite (300 mm)	3.8	33
Sandstone (300 mm)	2.5	44

* Thermal transmittance: (U, watts per square metre area per degree of temperature difference in Kelvin, same size unit as Celsius) the rate of heat flow through a whole wall construction. It is inversely proportional to R, resistance of the wall (or sum of resistances in a composite wall).

P: the theoretical time for a heat pulse to completely pass through a wall.

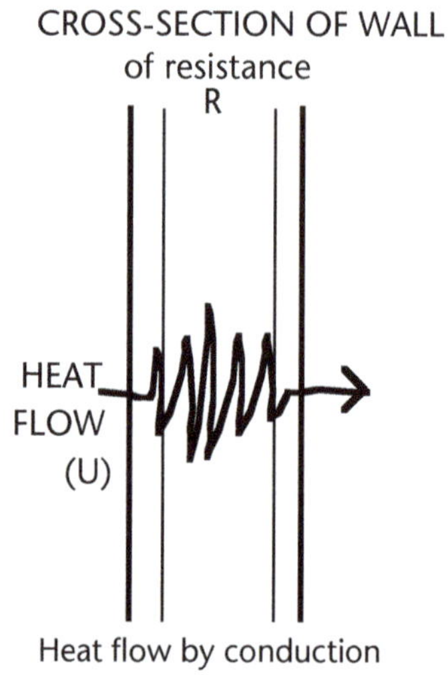

The basics of passive solar house design

Using thermal mass materials (including stone) is a basic part of passive solar house design. Another is the location of windows. It doesn't cost more to put the

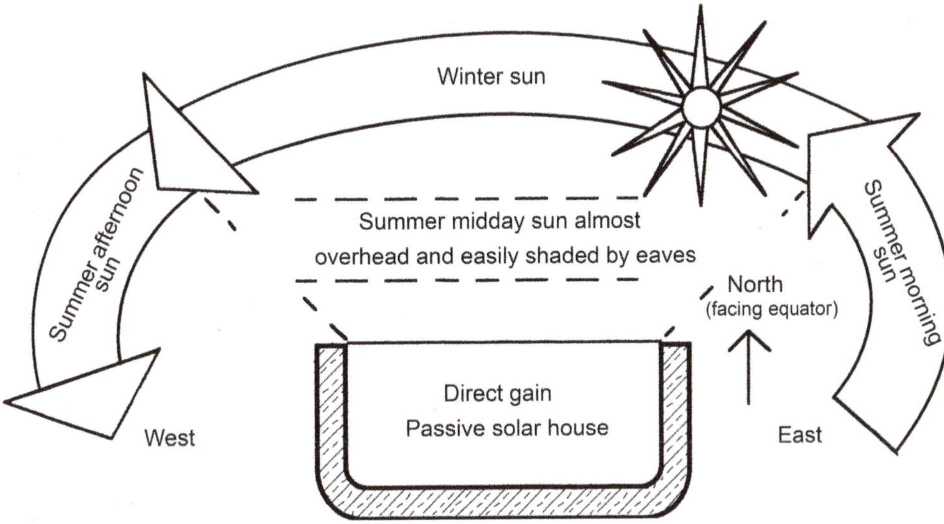

Figure 1.2: Path of the sun.

larger windows in a house so they face the sun in winter, i.e. as close as possible to due north in Australia (it is the opposite in the Northern Hemisphere). In summer, the sun goes higher overhead, and doesn't shine through north windows if there are overhanging eaves.

Figure 1.3: House with north-facing windows, near Bacchus Marsh, Victoria, built from local stone quarried on the property.

Figure 1.4: Curved pale stonework of internal walls, house near Bacchus Marsh.

In colder climates, the wall thickness alone may give enough shading over glazed openings. Windows on the east and west should be minimal. Windows on the south should provide adequate light and ventilation.

Where site conditions make a north aspect difficult, the rule of thumb for window orientation is that the limiting angles are 15° west of north and 30° east of due north. Solar rays have greater heat intensity in the afternoon, and sunlight is more desirable in the morning, to establish a comfortable indoor temperature for the day. Beyond these angles, sun control becomes more difficult.

In summer, especially in a climate with hot days followed by cool nights, the walls cool down at night (with the help of cross-ventilation), and stay cool during most of the day, absorbing heat from the air and from our bodies. Even with a hot air temperature the house feels comfortable because of the stone walls (see Figures 1.3 and 1.4). The right balance of solar gain through glass, thermal mass and insulation can substantially reduce heating and cooling requirements.

As for views; first, the view is often more enjoyable from the outside anyway, and a verandah or terrace can be placed to make the most of it.

If the best view is to the north, that's wonderful. Heat lost through north glass in winter is gained again when the sun shines. Sun penetration can be controlled with the right amount of overhead shading for windows, where they

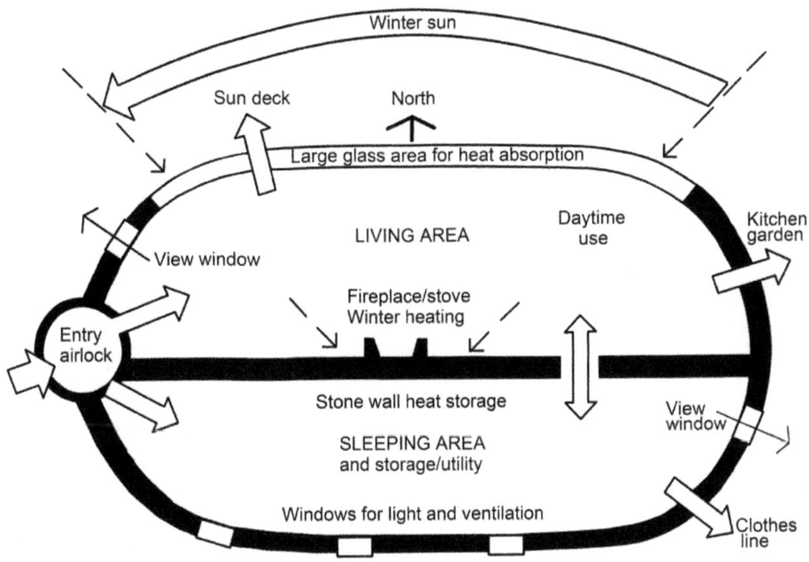

Figure 1.5: Concept diagram, the interactive stone house.

are wanted to enjoy the north view, and heat loss can be reduced with the use of insulating curtains.

However, spectacular views have a way of being to the east or west, where large windows would heat up the house too much in summer when the heat isn't wanted. To solve the problem of low angle sun penetration from the east or west, the first solution is often to put in blinds – but then people can't see the view anyway, which defeats the purpose of a window in that position. Heat gain is more effectively reduced by using smaller windows where direct sunlight is not wanted, but big enough just to give a taste of the quality of light outdoors (see Figure 1.5).

With windows to the south, unwanted summer sun is minimised, although some may come in at a low angle of incidence in the early morning or late afternoon. The main problem with south glazing is winter heat loss. More heat is transferred between inside and outside by windows than walls, particularly when there is a significant difference in temperatures. Unlike south glass, north glass (and good curtains) can produce a net heat gain from the balance of winter sun with heat loss.

With these things in mind, you can take advantage of some views while minimising windows to the east, west or south. A window can frame the view, like a picture to be seen from a favourite chair, or from the kitchen sink. Doors can line up with a window to borrow the view. An internal door can be closed off if conditions are extreme. An open-sided outdoor room can also achieve this degree of shading. Sometimes just a fleeting glimpse gives a welcome

Figure 1.6: Looking out to a verandah, this courtyard house near Bacchus Marsh is framed by round stone elements, which add character.

surprise, or lets it be known the sun is setting, so you can go outside to experience its full impact.

Stone is an ideal wall surface in rooms that collect direct sunlight through a large area of glass (see Figures 1.6 and 1.7). The stone wall acts as a heat equaliser, absorbing heat when the sun shines, so the room does not get uncomfortably hot, and storing the heat when the sun is not shining. A masonry finish on the floor also has this effect.

Stonework (or any masonry) placed where it can absorb direct sunlight can absorb heat about four times more efficiently than if it is in the shade (Figures 1.8 and 1.9). When solar radiation comes through glass into a room, some is absorbed and stored in building elements, some is reflected or re-radiated. Convection currents in the air circulate the heat to cooler areas. A space containing little or no thermal mass will heat up quickly when the sun shines in, but cannot retain this heat for long when the sunlight is gone. Insulation can slow down the heat loss (or gain) but it cannot store the heat.

Room layout

The location of rooms can be worked out according to how they will be used and when. Rooms used in the daytime should mostly be to the north to take best advantage of the sun, and those used overnight grouped elsewhere, with a dividing door between the two parts of the house (see Figures 1.5 and 1.10). Room

Figure 1.7: Interior kitchen wall, openings give view to courtyard.

Figure 1.8: Carport interior, courtyard house.

Figure 1.9: The tower at Narrapumelap, near Wickliffe, Victoria, built of basalt and limestone. The wall is covered in Boston ivy, which hangs on with suckers, better for the longevity of the wall than English ivy with its invasive roots.

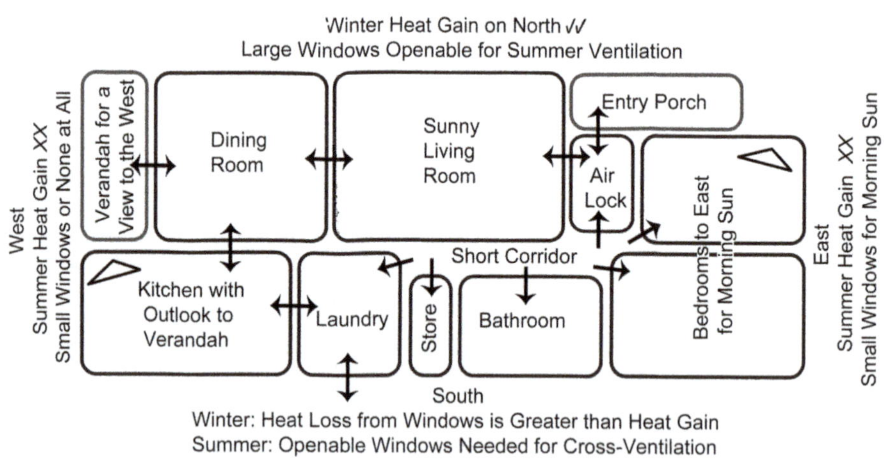

Figure 1.10: Flow between spaces in a house – design carefully to minimise waste space and use a checklist to keep track of the required spaces.

temperature in the daytime area can be higher than that needed in the sleeping area, so that closing the door between the two areas can reduce heating costs. At bedtime the door can be opened to let the heat flow to the sleeping area.

Daylighting

Internal stonework should preferably not be too dark in colour. It can make rooms very dim. With a dark stone such as basalt, it may be beneficial to increase the size of windows to compensate where it is used internally, and to contrast it with pale surfaces for the ceiling, floor and furnishings. If the dark stone is used only for a feature such as a fireplace, this is not such a great concern. In fact, most building stones are darker than the pastel paint colours usually used for interiors, so for those prepared to labour over internal stonework, think ahead on how to solve the problem of dimness. After all, there is little satisfaction in working painstakingly on face stonework if it will just end up under a coat of paint.

Window and door openings give expression to the stonework. The play of light is something that can be anticipated through the shape and placement of windows and glazed doors, or an occasional fanlight over a door. These openings may do more than just let light in.

Figure 1.11: Historic two-storey stone house with timber lintels, Yandoit, Victoria.

The transition from a rough stone exterior to a smooth angular interior can be compared with crystals on the inside of a rock (see Figures 1.15 and 1.16). With the use of face stone internally, the mood is different again, as the individual stones in the indoor scene add their own character.

Construction considerations

The house design can be worked out to simplify the process of building stone walls. The building design should respond to the potential of the available stone. When corner stones are scarce, it is worth considering rounded walls or brick corners. Some parts of a house are more important, and need to be done with quality stonework, but in other areas quicker construction techniques or materials such as formed stonework or timber framing can be used.

An awareness of the character of individual stones may direct changes in the treatment of walls or windows. This can add to the satisfaction that a handcrafted building gives. Easily worked sandstone deserves a more adventurous treatment, such as arches over openings. Large flat stones are good as lintels or for a mantelpiece.

If you plan to build in stone because of a philosophy of using locally available natural materials (Figures 1.13 and 1.14), you may want to avoid the use of reinforced concrete or steel in the building. In this case, door and window openings can be spanned by timber lintels. If the stone comes in suitable shapes, or can be worked with a chisel, stone lintels or arches are possible. Of course, if the stone is not suitable other options must be considered.

The local council may be concerned about the quality of the stone for arches and lintels. If the local stone does not lend itself to this use, think about how far you would have to transport a more workable stone, or the cost of having a quality stone cut to shape and brought to the site. These options could be prohibitively costly.

For renovation, the challenge is re-using an existing stone building. The damp-proofing may be ineffective, or there may only be small windows, things that reduce the quality of life. These failings are just as likely in a brick building, but it is easier to alter a brick building, and to update it. In an old stone building, any deviation from the original stonework is obvious. Efforts to modernise must be done with utmost care. Replacement of cracked or otherwise damaged stone with stone from a different source can raise issues of whether the reconstruction is authentic. If the fault lies in the detailing, is it right to replicate it? In historic buildings, people want to recognise the heritage value. In many cases it is best to leave the old intact, clearly different from the new. Building from the ground up is less constrained.

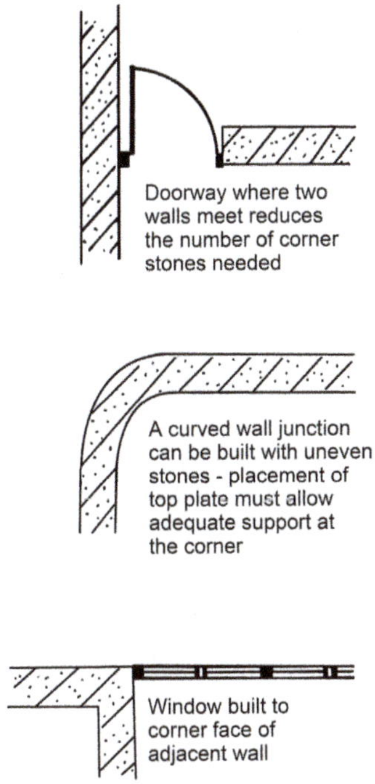

Figure 1.12: Corners are needed for stability but here are some ways to avoid unnecessary corners.

Thinking outside the square

Shaping corners and arch stones is time-consuming. The use of curves in a design instead of corners can reduce the number of corner stones, while giving a very pleasant effect.

Some stones naturally break with corners at a consistent angle other than 90°. Where there is a plentiful supply of these, a few corners can be designed at that angle without increasing the time spent on shaping.

Such features, like the use of curved walls, add interest to a house, and can be included so long as they do not create difficulties with construction above them, or the furniture layout. Corner stones are also needed for window and door surrounds. It is best to minimise the use of corners in the design, except if the stone is easily worked or has natural corners, or if the aesthetic effect is wanted, so the extra effort is well spent.

Likewise, narrow free-standing stone columns add to the cost. Unless there is easy access to an old stone building being demolished, or a friend who works in a commercial stone quarry, solid stone columns don't come cheaply. The construction of a square column from smaller stones needs four sets of corner

Figure 1.13: Arched windows facing north in tower, house near Hepburn Springs, Victoria. The kitchen window is the lower window.

Figure 1.14: Interior of round kitchen.

Figure 1.15: Castlecrag house contrasting smooth stone with rock-faced stone. The smooth faceted edges are used for the transition between inside and outside, a reference to the way crystals form inside a rough rock exterior.

Figure 1.16: Contrasting rough and smooth surfaces, Castlecrag, New South Wales. The neatly worked window detail varies the circle with the use of radiating crystalline forms.

Figure 1.17: Framed attic roof above rubble stonework, Delicknora, Victoria.

stones. Check the plans to see if all the corners are necessary, or whether there is an easier way to build any of them (Figure 1.12).

Some buildings use bricks for door and window surrounds and corners. They provide an effective contrast with a stone wall. Bricks are stronger than some kinds of stone, and easier to fit together than irregular stones.

Timber may be preferable over window and door openings if it is only supporting an upper floor or the roof. Timber columns and lintels are good in such places.

The use of stone above openings looks good, but the easiest and cheapest solution is to position windows high enough in the wall so that a lintel only has to support the ceiling and roof structure (see Figures 1.17 and 1.18). In old houses, timber lintels often support stonework as well. The timber used needs to be durable and very close to the width of the wall above (sometimes it is a pair of lintels). This is something that should be discussed with the local building surveyor. For spans of about a metre, a squared timber beam the width of the wall is probably enough. Only a small number of stones should be directly supported by the lintel, with the load being spread to the sides.

Arches over smaller windows can give a pleasant feeling to a house. Windows and frames to fit under a stone arch are non-standard, and are difficult to find second-hand. Openable arched windows are less available, but most window

Figure 1.18: The stone walls are only just higher than door height, Delicknora, Victoria.

Figure 1.19: Some window options: fixed pane for stone lintel, fixed fanlight for brick arch, specifically made double hung window with stone arch and timber lintel.

manufacturers can make fixed arched windows to order, for the top of a window or a doorway. An arched window can be combined with a rectangular frame. The window frame is placed on the inside of the wall, with a lintel for floor construction above it, while an arch on the outside supports the upper stone wall (see Figure 1.19). This arrangement works well for a two-storey building where the wall is narrower upstairs. It should be possible to find suitable stones for a rough arch, if the dimensions are not too large (the keystone, at the top, is the most important). The sides of an arch must allow for the sideways thrust created, so the location of the arch must not be at the end of a wall with only a narrow pier at the side. There should be some width and solidity in the stonework at each side.

To span large openings, the council may want assurance from an engineer. This cost should be allowed for if the design requires it. Steel or reinforced concrete lintels may be economical for large spans with stonework above. Careful detailing is needed to conceal the structure behind a stone face. This may involve matched widths of stones (or the use of bricks) to sit on a steel lintel. Veneer stones can be placed inside the formwork before a concrete lintel is poured.

When designing a stone house, the details should be considered, including availability of stone and other products, to avoid forcing last-minute compromises. Make sure there is a source of stone for corners and window surrounds. Establish where the doors and windows will come from, especially if arches are intended. Get prices on the undecided elements in the building before the details are made final.

Figure 1.20: Recycled basalt cobblestones used for east and west walls with timber-framed glazed walls on north and south, Daylesford, Victoria.

Figure 1.21: The courtyard, framed over with wire netting and view towards vine-covered chimney wall, house near Bacchus Marsh.

Cost-effective house design

Careful placement of windows, the use of massive stone walls, and roof insulation will reduce heating and cooling loads (and costs) for years to come (Figure 1.20). Extra time spent working out design problems may be tedious, but could mean the difference between a totally enjoyable house to live in, and one with the frustration of unused and unresolved areas.

The external structure creates the space within. A round building seems the most efficient shape, and the use of curves reduces the need for corner stones, which is a cost saving, but a round roof takes more thought and effort to build.

'Negative areas' such as courtyards must be carefully located and have a good use to justify the extra length of wall that does not enclose internal floor area (see Figures 1.21 and 1.22). These areas are subtracted when the floor area is calculated. Think about where to extend or make changes later.

While deterioration can be expected in other materials, where replacement and change may follow at some later date, stonework detailing should only require minimal rehabilitation unless it is in a very exposed location (Figures 1.23 to 1.27). For sills and other areas exposed to weathering, stone should be chosen for its

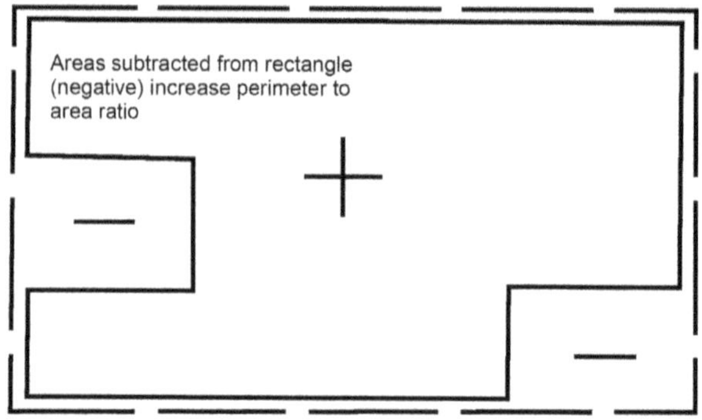

Figure 1.22: Negative areas.

resistance to water. Openings in stonework deserve careful consideration. A tall enough window may become a future door if there is a need to extend one day.

Make special allowances for plumbing fixtures and other services. If there are any doubts, or ideas beyond the present budget, leave small holes or ducts in walls for anticipated changes. Gaps can be filled with a weak mortar that can be easily

Figure 1.23: The grand façade of Narrapumelap, Wickliffe, Victoria, softened by lacework.

Figure 1.24: Next to the turret, decorative gothic windows, decorative buttresses and a matching turret window under the gable roof of Narrapumelap.

Figure 1.25: Carved in limestone, the lion's head keystone adds drama to the entry doors. The smooth-finished basalt corners are buttressed, giving stability to the wide arch, and a second basalt arch rises above the limestone from large springers. More limestone is used to arch the upper tower windows, but the turret windows are capped only.

Figure 1.26: The service wing is of simpler construction, all in basalt, Narrapumelap.

removed, to save time and effort in the future. If room widths are no more than 4 or 5 metres, the structural timber spanning the room can be kept to economical and readily available sizes. Work out the room sizes carefully. If you intend to build internal stone walls it is best to maximise room areas so as to minimise the number of internal walls.

Figure 1.27: Pantry ventilation opening, Narrapumelap.

Perimeter proportions

A longer north wall improves sun control via windows. Minimising the ratio of external wall perimeter to floor area reduces overall heat loss as well as building costs. To optimise both sun control and heat flows in a temperate climate, the house should have proportions of about 1.5 to 1 between a well-glazed north wall and the closed east and west walls (i.e. half as long again). In hotter climates the proportion should be less, since there is less demand for winter sunlight.

Check the dimensions of rooms similar in size to those you want to use. Make adjustments so that you are confident that the areas provided in the house plan will best serve immediate and future needs. It is much easier to change the plan on paper now than to move a solid stone wall later on.

Figure 2.1: Narrapumelap service entry, the grammar of basalt sill and limestone lintel is varied by the use of a sill as a lintel for the window below. Secondary keystones in the coursework over the openings indicate the structural purpose of providing two flat relieving arches.

> *'When rocks either dry from a moist state, or cool from a heated state, they necessarily alter in bulk; and cracks or open spaces form in them in all directions. These cracks must be filled up with solid matter, or the rock would eventually become a ruinous heap. So, sometimes by water, sometimes by vapour, sometimes nobody knows how, crystallisable matter is brought from somewhere and fastens itself in these open spaces, so as to bind the rock together again, with crystal cement. A vast quantity of hollows are formed in lavas by bubbles of gas, just as the holes are left in bread well baked. In process of time, these cavities are generally filled with various crystals. ...*
>
> *Mary: But where does the crystallising substance come from?*
> *L: Sometimes out of the rock itself, sometimes from below or above, through the veins.'*

John Ruskin (1865), *The Ethics of the Dust*

2

Geology of building stones

To understand building stone, it helps to know something of how it was formed. Stones vary in hardness and durability, and some are not suitable for building because they are too susceptible to weathering or break up too easily.

Different types of stone have different characteristics, according to the rock formations where they occur. They can be used in different ways to make the most of these physical properties. For example, the more durable stone can be used at the base of a building where it will be exposed to the weathering action of water, and softer stone can be carved into an arch under a protective roof where it will not be weathered away.

Rock can be classified into three main geological categories: igneous, sedimentary and metamorphic.

- Igneous rocks are those formed from the molten rock beneath the Earth's crust.
- Sedimentary rocks are built up from layers of silt deposit.
- Metamorphic rocks have been hardened by being exposed to high temperatures or pressures from adjacent rock formations.

Within these categories, stone types are further divided according to their chemical make-up and physical properties. There is so much variation in the geological characteristics of stone that it may be difficult to classify a stone exactly into one type or another. Plenty of stones display the properties of more than one geological division.

For the stoneworker it is more important to understand the physical characteristics of the stone being used, such as whether it is coarse or fine-grained, layered or crystalline, and in what direction the lines of weakness run.

Figure 2.2: Granite wall with cantilevered shelves, locally quarried, with bricks at openings to reduce stone shaping, near Moruya, New South Wales.

Building stone

Commonly used building stones are those that can be worked with a chisel, such as sandstone, limestone, granite and others with similar physical properties. Denser stones such as slate and marble are used for flooring, veneers and finishes. Stones such as slate have layers, and can split neatly into flatter stones, but they are difficult to cut across the grain without a specialist tool such as a diamond-tipped power saw.

Stones that are brittle due to weathering may be available locally as field stones or river rocks in sizes that are easy to handle. If they are found near the building site it is worthwhile to build with them. However, very rounded stones have inadequate top and bottom bedding surfaces, so will not sit in place firmly, making them undesirable for structural walls. It may not be feasible to shape them with hand tools, except to knock off small bits from the edges to fit them better into a wall. Even this may be difficult, as hitting some stones too hard can shatter them into useless pieces.

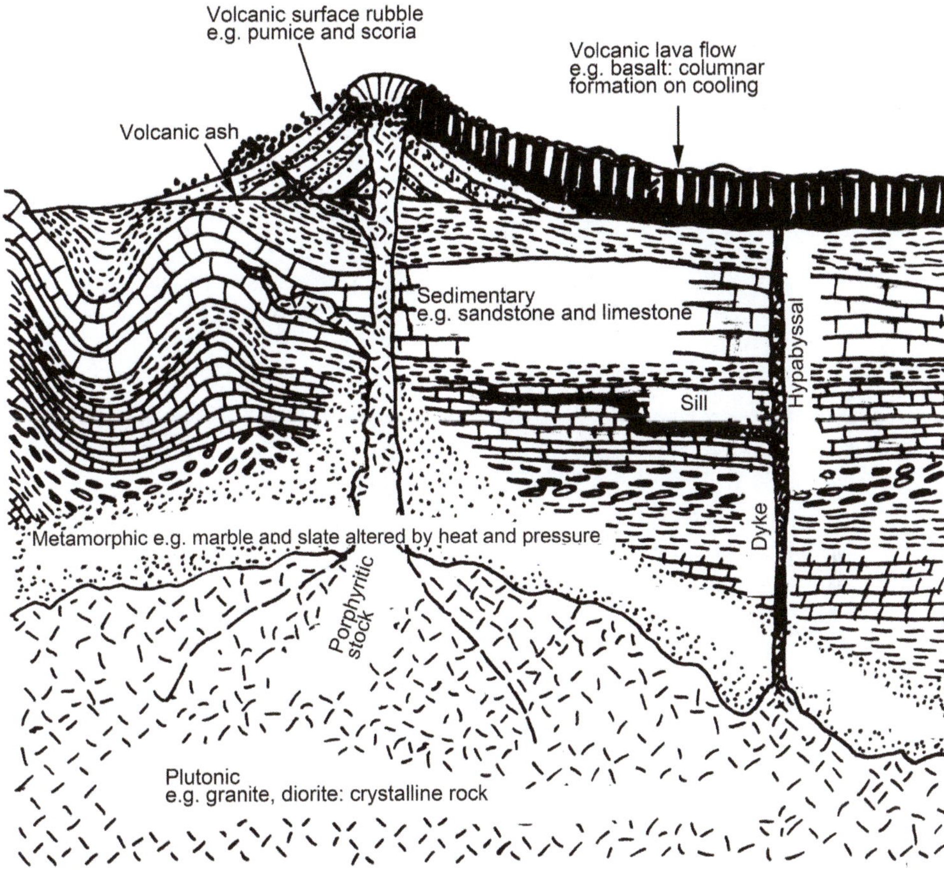

Figure 2.3: Cross-sections of rock formations.

Igneous rock

The characteristics of igneous rock largely depend on whether it was formed above or below the Earth's surface, and how long it took to cool down, as well as the chemical composition of the original hot magma (or molten rock) from which it was formed.

Volcanic rocks have formed in a lava flow, and cooled relatively quickly. Those lying on the surface are often rounded or odd-shaped and very bubbly (Figure 2.5). Basalt is the commonest volcanic stone used for building. Deeper lava flows form shrinkage cracks as the molten rock cools, and the resulting stone formation is shaped in vertical prismatic blocks, roughly hexagonal in cross-section. This is called columnar basalt (see Figure 2.4).

Figure 2.4: Columnar basalt formation in an old quarry, Daylesford, Victoria.

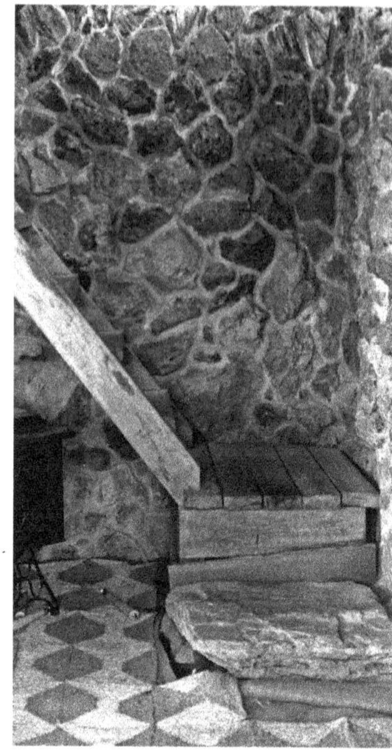

Figure 2.5: Owner-built walls of basalt rubble at Yandoit, Victoria. At the base of steps to a loft, a flat sedimentary stone is more suitable.

Plutonic rock, such as granite, has formed deep below the Earth's surface (remember Pluto, the Greek god of the Underworld?) and taken a long time to cool. It has gradually formed into crystals of different compositions. In granite, these crystals are predominantly quartz, mica and feldspar (which gives the stone its specks of colour) (see Figure 2.6). Plutonic rocks are generally very old, as the softer rock that covered them when they were formed would have taken a long time to erode away.

Hypabyssal rocks are partly crystalline, formed where magma has intruded into cracks in existing rock formations (Figures 2.7 and 2.8). These rocks have cooled faster than plutonic rock but slower than volcanic rock. Magma that cools quickly becomes glassy and structureless, but where it is slower cooling it forms crystals. Hypabyssal rock can form a kind of glazed surface in the cracks between sedimentary rocks. Where cracks are wide, larger formations of quartz and porphyritic crystals are found, formed during slow cooling on the way to the surface. Where the rock is between bedding layers it forms a horizontal sheet, called a sill, when vertical it is called a dyke. These sheets are not usually very thick, so rocks that form this way are generally smaller than volcanic or plutonic

Figure 2.6: Eroded granite boulders at Wilsons Promontory National Park, Victoria, showing crystalline structure.

Table 2.1: Igneous rock

IGNEOUS ROCK	BASIC	INTERMEDIATE		ACID	
	45 to 55% silica	55 to 65% silica		Min. 66% silica	
Mineral	Olivine	Iron ores		Biotite (micas)	Muscovite
Composition	Augite	Feldspar	Hornblende	Quartz	
	Silicates of Ca, Mg, Fe, Al	Silicates of Ca, Fe, Mg		Silica	
Formation				Pumice	
– Volcanic	Basalt	Andesite	Trachite	Rhyolite	Obsidian
– (fine grain)					
– Hypabyssal	Dolerite	Porphyry	Felsite	Quartz porphyry	
– (medium)		(microdiorite)			
– Plutonic	Gabbro	Diorite	Syenite	Granite	
– (coarse grain)					

Figure 2.7: Pink granite formed alongside grey, Eurobodalla National Park, New South Wales.

Figure 2.8: Dark volcanic rock filling a shaft in granite, Eurobodalla National Park.

rock. Dolerite resembles basalt but is coarser grained, and is more often used for road stone than building.

To help with comprehension of the geology of igneous rocks, Table 2.1 outlines the variation in types. Some of these rock types are not commonly used in building.

Basalt that has formed in a columnar formation often breaks away into blocks that can be handled by one person, making it a relatively easy stone to collect. It is a dark, hard, fine-grained stone. Basalt river rocks can have a yellowish-brown weathered surface and rounded edges. They may be difficult to distinguish from the surrounding sedimentary rocks. Although basalt is a very hard rock, it can be worked because of its fine grain. Small pieces of the stone can be broken off neatly, if hit hard enough with a chisel.

Surface rock may be bubbly and rounded. In some areas dry stone walls have been built from these, to clear paddocks for farming. If it has cooled very quickly, it is of a different colour and consistency from basalt and is referred to as scoria.

Granite and granitic stones (porphyry/syenite/trachyte) have a speckled, coarse, grainy texture. Because they are durable and can be shaped, granites are used at the base of buildings and for steps and thresholds. Granite polishes well and is used as an ornamental stone, especially when it contains red feldspars, which give the stone a rich colour.

Other igneous rocks are used in the immediate area where they are quarried, but are generally not very easily worked.

Sedimentary rock

Sedimentary rock is made up of layers that formed from sand, silt and clay deposits as a result of earlier rocks eroding. These bedding layers have been compacted by the weight of more sediment above, and over time they may have been folded and uplifted by forces in the Earth's crust. The depth of these layers (or strata) and their grain size will determine how suitable they are for building.

A sedimentary rock formation may consist of a repetitive series of different layers, formed by different sized particles in a series of deposits (e.g. sandstone, shale). Or it may be very consistent, from one kind of sediment deposited in a large quantity.

Sandstone is made up of tightly compacted sand particles, usually fairly consistent in size (see Figures 2.9 and 2.10). The colour may vary to indicate the bedding layers. A common example of sandstone formation by sediment deposit occurs at a river mouth. The river drops sediment in its valley and flood plain as it flows, and it usually carries a great deal of suspended mud and sand (especially during floods) when it reaches the sea or a still lake. Here it is slowed down, mixing with a large body of water. This causes the coarser particles to settle quickly, while the mud flows further out before settling.

When a river is in flood, the sediment may be spread over a large area before settling in a thin layer. The tributaries to a river sometimes bring down sediment of different colours or composition. If these flood separately, the successive layers are of different colours and textures. Wind can also play a part in distributing sand or dust particles over a large area.

The consistent size of the particles in sandstone makes it one of the most versatile and easily worked stones. The strength of different sandstones varies considerably. Some are harder to chip with a chisel and others may be almost crumbly. Sandstone can be susceptible to weathering. After some time the softer layers erode away. It is worth looking at old buildings in a region, to see how the local stone has withstood weathering. Stone plinths and walls at ground level may be worn away by water, and stone parapets, sills and decorative work may have deteriorated.

The minerals that make up the grains in sandstone are predominantly quartz, and may also include mica, feldspar, dolomite and iron compounds. In water, quartz and mica settle at different rates, so in a stone the two are roughly segregated along close bedding planes.

Figure 2.9: Disused sandstone quarry at Darley, near Bacchus Marsh, Victoria.

Figure 2.10: Sandstone standing rock, Gariwerd, Victoria. The sedimentary layers are somehow protected by a resilient cap, giving the formation an almost animal presence. As well as changing the colour, surface oxidation can harden the rock.

If a micaceous sandstone can be split along the mica planes, the stone is called a flagstone. When the particles settle, they lie flat so the rock that results will split most easily along the bedding plane. A sedimentary rock with more than 25% feldspar is an arkose, and it has formed soon after the erosion of granite, before the feldspar was dissolved into clay. Sand particles in the stone are cemented together by other minerals, which are sometimes water soluble. Sandstone cemented by iron compounds are coloured accordingly, the red oxide or the brown hydrated oxide being common.

Limestone has formed on the seabed beyond where the sediment from erosion of land has been carried. The deposits in the still water further out to sea are formed from the remains of marine organisms, shells, corals and seaweeds, with

Figure 2.11: Weathering reveals layers in stone that was originally laid with a worked surface, Dubbo, New South Wales.

occasional layers of clay from volcanic dust and decomposed pumice. Sometimes these substances are carried by the sea in solution.

The minerals in limestone are mostly carbonates of calcium or silica. The commonest is calcite (calcium carbonate). Many limestones also contain dolomite, sand or clay, bits of carbonaceous matter or iron oxides adding colour to the otherwise white or grey stone (see Figures 2.11 and 2.12).

Table 2.2 shows different sedimentary rock types.

Figure 2.12: Erosion of the base course in a wall of cut limestone.

Table 2.2: Sedimentary rock types

Rudaceous	Rubbly, large fragments of older rock	Conglomerate, Brecchia
Arenaceous	Sandy in grain and texture	Sandstone, Gritstone, Arkose, (Quartzite)
Argillaceous	Clayey, fine particles	Shale
Calcareous	Largely calcium/magnesium carbonate	Limestone, Oolite, Tufa, Dolomite
Other: Organic	Derived from plant materials	Coal, lignite
Evaporites	Dehydrated soluble material	Rocksalt, gypsum
Pyroclastic	Volcanic composite	

Metamorphic rock

Heat or pressure can harden a rock formation and create a different structure that tends to be more durable. The resultant rock is termed metamorphic. Commonly used examples are marble and slate (see Figures 2.13 and 2.14).

Slate is derived from fine-grained sedimentary rock such as shale or mudstone. When the sediment was deposited, the longer particles gradually settled flat, aligning themselves parallel to the bedding surface as the weight of sediment above

Figure 2.13: Stone known as slate from Taradale, used for rubble walls, flat and on edge, as well as for sills and paving, near Sunbury, Victoria.

Figure 2.14: Taradale stone formed into a rubble arch. Recommended practice is to place bedding layers perpendicular to the face and the curve across the arch, flat on the walls. In a protected location such as this, withstanding the weather is not essential.

1)
2)
3)
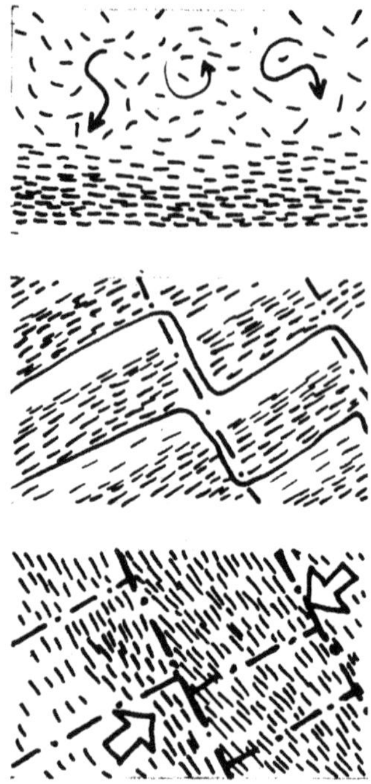

Figure 2.15: How slate is formed: 1) sediment settles, 2) layers in folds, 3) high pressure makes new layer with particles realigned.

pushed them down. During earth movement the rocks are subjected to lateral pressure, and shift, readjusting themselves into folds. When folding occurs, the beds of rock can move against each other, wearing away edges and forming joint planes. When the rock is subjected to extreme crushing pressure, the flat particles within the rock are squeezed together and realign themselves perpendicular to the pressure thrust. This process forms slate. The direction of cleavage in slate does not relate to the bedding planes (see Figure 2.15).

Marble is formed from granular limestone, recrystallised by heat or pressure. Impure limestones form crystals of augite, garnet or other minerals in the marble. Finer-grained stones are carved for statues. Some hard, fine-grained and oolitic limestones are used this way and may be referred to as marble. Coarser-grained marble is more generally used for building, though usually as a veneer wall facing or for decorative work such as fireplaces. Marble is not recommended for kitchen benches because it corrodes through reaction with acid (such as vinegar or lemon).

Travertine is a limestone, usually porous and formed from geothermal spring waters.

Serpentine is similar to marble but softer due to low-level metamorphic conditions. Usually greenish due to the presence of magnesium, serpentine is less susceptible than marble to corrosion by acid.

Some jades are of a serpentine type, others appear in gneiss.

Soapstone has been formed by infiltration of fluids through a serpentine-dominated rock mass. It is rich in talc, with varying amounts of other minerals, especially magnesite. It is very dense, yet easy to carve, and exhibits good heat retention.

Quartzite is a rock with some sedimentary qualities, resulting from the heating and cooling of sandstone that has not quite melted into magma. The quartz grains have grown larger and new quartz has formed, cementing the grains closely with silica. Quartzite is also the term used to describe a sedimentary stone made up of quartz grains that have been cemented at high temperature by a solution of water containing silica.

Gneiss has almost the same mineral composition as granite. It has partly recrystallised from the action of heat, but has kept some indication of its sedimentary layers. Some gneisses correspond in mineral content with syenite, diorite or gabbro. Although gneiss is as hard as granite, its other physical properties differ, and it is not as easily worked.

Schist is a distinctly layered rock of various mineral composition that has recrystallised. Schists and gneisses are often in beds that have been folded and contorted by earth movement.

Qualities of building stone

The behaviour of a stone when worked responds to its geological characteristics. It can also depend on the patience of the stoneworker. Stone is a natural material. When found in its natural state it can be used frugally or carelessly, depending on the user.

Every natural place has its own mood, or feeling. It is good to respond to this. Taking time to contemplate and consider how stone has been formed in its natural condition can help to understand the material itself a little better.

Within the many categories of stone, there are also regional characteristics, which can be learnt through observation or picked up from local knowledge. For example, basalt is called bluestone in Melbourne, but in other places it may not look as blue, or it may be brown. Local variation can occur over short distances. Clues of quality stone nearby may be found in rock faces at creeks or riverbanks, as well as road cuttings or rock outcrops. These geological formations usually indicate that similar stone can be found nearby. But alternatively they may indicate that the local council has used stone from somewhere else, perhaps to stabilise against erosion.

Stones may weather differently, with no clues from their other visible qualities. Some stones used for construction may be suitable for use in sheltered locations, or may stand the test of time if finished with a rough rock face, but not with a smooth face (see Figures 2.16 to 2.18). Table 2.3 shows the building qualities of different stone types.

Figure 2.16: Base courses and underfloor access, worked to varying degrees of smoothness, Hunters Hill, New South Wales.

Figure 2.17: Stone walls and a chimney rising from a tiled roof, extended for the windows, Narrandera, New South Wales.

Figure 2.18: Long stone lintel used over doors, Narrandera. Note the two long stones in the next course spreading the load.

Table 2.3: Stone types – building qualities

	Av. density (kg/m^3)	Strength	Water resistance	Colour	Notes
IGNEOUS					
Granite	2580	Good	Good	Pink/grey/brown/black	Regional
Diorite		Good	Good	Pink/grey/brown/black	Very hard to work
Gneiss	2690	V. good	Good	Pink/grey/brown/black	Good wall stones
Basalt	2200	V. good	V. good	Black/grey	Dark, good for building
Obsidian		Good	V. good	Black/dark	Glassy
SEDIMENTARY					
Dolomite		Good	Fair	Yellowish	
Limestone	2500	Good/poor	Poor	Yellow/grey/white	Sometimes chalky
Shale	2670	Poor	Poor	Grey/brown	Soft, good surface
Sandstone	2200	Good/poor	Fair	Red/yellow/white/grey	Good wall stone
METAMORPHIC					
Slate	2790	Good	V.good	Grey/black/green	Seldom best quality
Schist		Good	Good	Sparkled	Good surface
Marble	2700	Good	V. good	Red/pink/white/grey/black	Decorative
Serpentine		Fair	Good	Green	Regional, ornamental

Stone for heritage work

Old buildings may show defects in the local stone over time, in much the same way as does weathering in the natural environment. In heritage work this can present a dilemma – is it best to replace defective stone with stone from the same source, for authenticity, or to use a stone from somewhere else that is known to be more durable?

As a result of urban growth or the location of a former quarry in a conservation area, it may not be possible to go back to the same source. The same stone may no longer be available – commercially or otherwise. If it is not available, it seems extreme to source a replacement stone from so far away that it is prohibitively expensive, just to find a match. The solution should seem reasonable, yet market forces have changed and many imported stones are now available that would not have been contemplated in the past. One might ask: what would the original builders have decided on, if faced with this problem?

Figure 2.19: Historic sandstone water filter, Sydney, New South Wales.

The geological make-up of stone for building determines its workability, weathering characteristics, strength, colour and porosity (Figure 2.19). All these weigh up together for an assessment of the available local stone. For more precise stonework, a stone with varying geological characteristics may be brought in. The advantages may include workability, colour contrast, natural shape or commercial availability.

Figure 3.1: Bedroom alcove with built-in desk unit, house at Bacchus Marsh, Victoria, in local stone.

'... You cross a little lake, bridged at one point by a single slab of rock; then up a ravine, down which a veritable mountain brook is tumbling, and through a rock foundation so natural, that, until a series of faults and dislocations, synclinals and anticlinals, in rapid succession arouse your geological memories with a rude shock, you cannot believe that all this colossal mass of material has been transported here by man, from distances to be measured by leagues; and that a few hundred years ago a low plain existed where now are rocky ravines and dark dells.'

Edward S. Morse (1886), *Japanese Homes and their Surroundings*

3

Siting and stone availability

Stone is a naturally occurring building material, abundant in some locations. To build with stone, it makes sense to collect or quarry suitable local stone, rather than hauling it from somewhere else to the site. The cost of building in stone can vary a lot, especially if it has to be transported to the site. While in some locations stone is scarce, in others it is lying around waiting to be collected and used.

The quality of the stone is another consideration. A building site may have some kind of stone available within a reasonable distance, but it is often difficult to shape. It is more effort to build strong walls with rounded or irregular-shaped rocks. Blocks of stone that are too big to be handled, and cannot easily be split into smaller sizes, are too difficult to get to the wall. If the stones are too small, the wall gets built slowly, as there is more stone handling. Each layer, or course of stones, must set into place before the next, and more mortar is used for smaller stones.

When building in stone, it is best if the stone comes from reasonably close to the site. The cost of haulage is prohibitive over long distances. Bringing stones from a long way off is only feasible for a small part of a house on a normal budget. If stone is to be shaped by a stonemason, it may be cost-effective to purchase a more easily worked stone. In this case, the stonemason can advise on stone supply. If the source is close, stone may be cheaper than other wall materials. Collecting a few outside stones as memorabilia from one's travels is a likely scenario, but this will not get walls built in a hurry without a supply of local material as well.

The blocks of stone to be used, whatever the local stone type, come naturally in a variety of shapes and sizes. A wall built from rough-shaped stones is a rubble stone wall, and is built differently from an ashlar wall, which uses sawn or accurately worked stone. Building a rubble stone wall involves constantly having to

Figure 3.2: Barn walls, built in natural rubble. Corner stones are of irregular height, yet have been carefully chosen to finish square. House ruin in background. Near Hepburn Springs, Victoria.

choose which stone will fit best (see Figures 3.2 and 3.3). They do not all meet at the same angle, unlike bricks or concrete blocks. It is a bit like building a jigsaw puzzle. And, like a puzzle, it cannot be finished until all the pieces are there.

For this reason, it is best to collect as many stones as possible around the site, to give the widest variety to choose from. Unlike a jigsaw puzzle, if a piece is not quite the right shape, it can be altered to fit better. Still, it saves time if the right stone can be found to fit without needing any edges broken off.

For building straight wall sections, almost any kind of rock will do, but rougher stones take more time to fit together. A wall can be built more easily with stone that is able to be shaped with a hammer and chisel, so any protruding lumps are removed. With suitable types, it should be possible to trim the rough edges to make the stones fit closer together. Even this is difficult with some stone, so the difference has to be made up with mortar.

The choice of stone

Before making the commitment to build in stone, it helps to know the limitations and potential of the local variety (Figure 3.8). It is nice to fantasise about the house you want to build, but not always as easy to get the right stone to build it. There is an advantage in being familiar with the type that will be used. However, it is common enough to learn on the job.

Figure 3.3: House ruin showing smoothed off internal walling and timber lintel, near Hepburn Springs.

Factors to consider in design are colour, wall texture, light and dark (especially in interior face work), hardness and resistance to weathering. You may choose to mix different types of stone to avoid the dominance of one colour, or to build straight wall sections in one type, and corners and window surrounds in another.

Figure 3.4: Rubble walling with squared corners, Molong, New South Wales. Flat arches relieve the loading of timber lintels.

The time it takes to gather enough stones for a house cannot be ignored. Unless the stone can be delivered from a quarry on demand, many hours should be anticipated fossicking through the likely stone source area, whether it be a creek bed, an old quarry or mine tailings. If there will be special features in stonework for the house, it might take days, and a trip to a further away location, looking for just the right set of stones. This can include chasing up commercial stone supplies if the building requirements for any of the work cannot be properly met using the local variety.

The site

Even if there is stone on your own land, there still may be a lot of work in quarrying it or breaking it into suitably sized blocks. By the time that is done, you will most likely have a pile of reject rocks to bury in the quarry pit or to use around the garden. The value of the natural stone in the landscape is another consideration, and any damage done quarrying will need to be rectified in time. Whether collecting the stone from your own land or elsewhere, you will probably need a truck or ute to haul it to the house site. With the ideal site you can get by with a wheelbarrow.

It is all very well to have the resolve to build in stone, but it is also wise to select a site where the local stone is abundant. This gets around the cost and trouble of transporting stone over any distance.

Once a suitable piece of land is found, the exact position of the house must be determined. On sloping land (common enough where there is stone) the places where the contours flatten out are already the best suited to outdoor use. Why would you want to lose those? Areas near the top of a hill can be utilised if possible as vantage points in the landscape, where not obstructing the best view corridors. More rock outcrops are usually found higher up. The house site should be below the crest of a hill to avoid being too windy. Low-lying fertile land is better used for growing food. The house should not encroach on the best garden area, so it should be built up-slope, where the view is better. Where waste is disposed of on site, there is an added advantage in letting go of grey water as high as possible into the water table.

Some sites yield good building stone when excavated for construction. This is a good start, but more will usually be needed. If good stone is not actually found on site, but can be collected nearby, further investigation is needed.

The most obvious indication of a good local building stone is the existence of stone houses already in the neighbourhood. The more there are, and the more consistent the stone in their walls, the better the chance is of there being a similar stone near your own site (Figures 3.4 and 3.5).

However, such things should not be taken for granted, so check for yourself where the stone for any such buildings has come from. Well-established locals

Figure 3.5: Large blocks of stone, arched at door, smoothed for lintels and end windows, margins worked at corners, parapet indicates durability, Castle Hill, New South Wales.

should know of any quarries. Find out where they sourced the stone that they used to build their own and any other houses, and try to find the quarry if there is one. Sometimes people who sound authoritative have never checked for themselves whether all the information they impart is correct. When a supply of stone is found, think about how many truckloads could be picked up, and whether that would be enough for the amount of stonework required to build a house.

There may be more than one type of stone in the region, which gives you a choice of which to build with. Different types of stone can be mixed in the wall, for different purposes. If it is more durable, it can be used for a base course or for sills. Or it may be better to make do with the stone that is easiest to collect.

Variety in stone or other materials

If there is no stone locally that can be shaped, consider other ways of finishing off the corners. With luck, enough naturally shaped corners could be found. There may be matching sets of stone with angled corners other than at right angles, and these can be used if the same angle is designed into the plans. It is possible that one of the stone types is workable with a chisel, and one is not. In this case, the workable stone can be used for corners and at door and window openings, while the rough stone is suitable for straight runs of wall.

If there are plenty of irregular stones but none that are easily worked into corner stones, there is the dilemma of whether to build very rough corners, or to use another material such as bricks. Buying a load of quality building stone may prove expensive, depending on how far away it was quarried. If you travel often to a place where good stone is available, it may be worth loading a few stones of the right size and shape into the car boot each time. Your fellow travellers are bound to act as if you are rather eccentric, but they will put up with it as long as it is legal. Do not collect in national parks, where it is an offence to remove anything natural from the environment (Figure 3.6).

It is possible to build up a small supply of good workable stones by collecting, enough for putting special touches in a few places, and using rough stone everywhere else.

The best way to resolve the problem of supply depends on how easy it is to get stone locally. Inadequate supply means the quantity of stone is limited, and the difference has to be made up with other materials. For rough stone, wide timber sections can be framed around windows and doorways, then the stonework packed up to the frames. Timber posts can also be used to support the roof if the stone is too rough-shaped to use structurally (or if you are in a hurry) (Figure 3.7).

Figure 3.6: Trial Bay Gaol, built of hard granite, sits above a former quarry in Hat Head National Park, New South Wales. This stone resource is no longer available, but granite also occurs on private land.

Figure 3.7: Granite rubble infill with post and beam construction, near Bruthen, Victoria. Roof construction before walls allows all-weather work, but posts must be durable.

Where there is a choice of stones, colour and hardness are the main deciding factors on how to use them. The warmer colours such as yellows and reds in a stone wall give a pleasant effect, but these colours are rare in some areas. Grey granite or

Figure 3.8: Historic stone wall with cracking, Merriwa, New South Wales. Cracks can indicate foundation settlement. Other factors are construction detailing and, where cracks go through stone instead of following mortar joints, a low-strength stone used with stronger mortar. If windows are not aligned, the distribution of loading is likely to be uneven.

Figure 3.9: Original flagstone external corner built in the 1920s, now enclosed by an addition, near Bairnsdale, Victoria. Pale stone works best for interiors.

limestone looks homogeneous, much the same colour as the mortar in between, so it might add interest to strategically place a few stones of different colours for some contrast. The odd piece of pink granite could be fitted in, for example, with grey granite. The mortar colour can also be changed to show off the stone more effectively.

Sometimes stones are cut to such regular sizes that they look similar to a concrete block wall when laid. However, stone is a sculptural material and can be shaped to add interest in a way that concrete blocks cannot. When stones are collected from a quarry, those of the right size and shape to be suitable for features such as lintels over windows, shelves and niches, arches or columns, are an advantage to have stockpiled. The diversity of stone sizes and shapes in a wall can make up for what is lacking in the colour variation of the stone. Building ledges, shelves and niches require some forethought as to how a room will be used. Shelves should be placed where they are neither at bump-head height, nor in the way of furniture.

The hardness of the local stone will control how easy it is to work it to the desired effect, and how long that will take. Stones such as granite and basalt, although hard, can be trimmed with hand tools with a bit of perseverance, as they have a crystalline structure that can be broken in any direction. Sedimentary stones (Figure 3.9) often have a tendency to break only in predetermined directions, either in their natural bedding planes or in joints or planes of cleavage

Figure 3.10: Stones with natural angle as found, matched for window reveals, near Hepburn Springs. This saves work squaring off corners, and lets in more light.

roughly perpendicular to bedding. They may shatter if an attempt is made to shape them in another direction. Some sedimentary stones are grainy and easier to work, such as sandstone and limestone.

If a good building stone is available locally, it is likely to be visible among historic buildings, especially in churches and public buildings. How well the stone withstands weathering can be seen by looking at the condition of the old stonework in these buildings, especially close to the ground, and on parapets or decorative work. If the old churches are all brick, it suggests the early builders didn't think much of the stone.

For the stoneworker using hand tools, stones can be classed according to their workability. Good building stones are those that can be chiselled to the required shape. Stones that are too brittle or too strongly layered to be easily worked can be built into rubble stone walls, but not used for corners and other trimmings that require a particular shape, unless they happen to be that shape already (Figure 3.10).

Table 3.1 at the end of the chapter lists approximate locations of some of Australia's principal building stones. The locations go back to old quarrying information. Some of these quarries are no longer in operation, so it is only a rough guide. In some cases the accessible supply of good stone is exhausted. In some instances quarries have good stone, but are controlled by companies or authorities

Figure 3.11: Old rubble wall with large corner stones, most of which have been plundered, South Australia.

that will not supply stone to the public. Not all the locations in the list have been visibly quarried. In spite of these difficulties, it may still be possible to collect some good building stone in the same region, but at other sites.

Occasionally good building stone becomes available through the demolition of an old house, barn or bridge, or a landowner wanting the removal of a ruin when it becomes unsafe (perhaps sometimes as a result of vandalism and stone theft). The stones in old buildings have often been shaped, and may be of better quality than others collected elsewhere. However, this should not be enough to justify the wilful destruction of old buildings to supply stone for the new (Figure 3.11). An old building may be suitable for restoration, and even a ruin can provide shelter for stock, as well as having a significant heritage value.

Stone may also become available second-hand through the replacement of weathered stone in public buildings or churches. This stone may be of a quality that is not readily available locally, since stone for important buildings is often brought from a long distance at a cost unthinkable for domestic work. Although the quantity of stone available second-hand is usually small, the quality can make it worthwhile to include some detail in the house design that would be difficult to build using just the predominant local stone.

Where to quarry

In a thorough search for the right stone at the right distance from the site, you may come across a disused quarry on private land that the owner will let you use (perhaps at a cost). Or you may find evidence of a formation that shows the promise of supplying quality stone, but has not been quarried before. If there are enough rocks that will break away evenly along straight cracks, it is possible to get them by hand quarrying.

The presence of good square-edged rocks on the surface in a rocky reef is an indication that more stones of a workable shape can be found by digging below. There may be a seam of quality stone in that location that is better than the surrounding stone. So resist the temptation to cream off all the good rocks from the surface, and instead leave them as markers to show where to dig for the good stone when it is needed.

Access may not be easy. It must be possible to get a truck close enough to load up, to avoid wearing yourself out carrying rocks back and forth. Walking along creek beds looking for stones is enjoyable, but carrying them back to the road is not as much fun. You will soon decide which stones are worth it.

Disused mine tailings and old quarries are often on private land, or else their access roads go through private land. Someone may have a lease to a quarry even where it is on public land. Respect other people's rights and do your best to seek out and talk to the person who regards a quarry as their own. They may intend to use rocks of a particular size or shape, but would be happy for you to browse through the others. This might help to avoid the conflict that is bound to occur if you are caught removing stones that they were thinking of using themselves. When somebody has built a road into a source of good stone, respect the work that has gone into it, and into establishing the quarry. You may even find a supply of stone that makes it worthwhile for you to build a road in there yourself.

The characteristics of a stone

Do not be too easily discouraged if the local stone will not be worked easily enough to be shaped at will. If there is plenty to choose from, look out for stones that are already close to the shape required and put them aside from the rest. By finding the right shaped stones instead of shaping them, you can save time, as chisel work can be very slow. Even when there is a need to shape a stone, it helps to look for one that is close to the required size already, and will not take too much chiselling into the right shape.

A house that uses rough field stone or river stone can have the aesthetic advantage of looking as if it grew naturally in its environment. Especially where natural rock formations are left intact around it, a stone house blends in with the colour and texture of its surroundings.

On sites with a plentiful supply of stones, there will be some that cannot be broken up easily into building blocks. These stones are ideal for use in landscaping, and can be saved for the garden. Some stones are weathered into interesting shapes, usually the wrong shape to make good building stones. Some are weathered and

Figure 3.12: Bricks used instead of stone for corners, over windows and for bonder courses at base of gable, Adelaide, South Australia.

liable to cracking. Some are covered in moss and lichen, which is too nice to clean off just so the stone can be used in a wall.

If there is not a sufficient supply of stone close at hand to select some that are the right shape for corners and so on, a decision must be made about whether it is possible to build the whole house out of the local stone. Stone brought in from elsewhere or delivered from a quarry may make all the difference. It is important to check the quality of the stone first. There may be only enough stone to build an outside veneer, together with brickwork, blockwork, formed concrete or timber framing on the inside (Figure 3.12).

It may be a cheaper and quicker alternative to use some bricks or timber. Bricks are good for corners, or internal wall faces (which may be bagged). If the stonework is rough, bricks can also be used for an upper wall bond beam. Timber framing can be used for internal and upper walls, without losing the benefit of the outside stone walls at ground level.

Not having a freestone to work with, but instead making do with odd angles and broken-off ends of stone, our experience was that we sought out better stone from many places. The quarries we found were not always being used to get stone, they were sometimes supplying crushed rock or other materials. Having maps with quarries on them was a good start, making sure these were in locations where stone was listed as suitable for building. Whatever the stone type, or the locality, the rest

Figure 3.13: Flat 'slate' from Taradale, Victoria, keystone set over fireplace as a feature, supported by steel lintel.

of it is exploration. Each stone is individual. Some more than others. Some are good for wall stones, others can make a feature (Figure 3.13).

Quarries come and go, depending on commercial fortunes, but there are usually other options still available when the quarrying business is gone. For example, a couple purchased local stone for their first house, but it was no longer available when they wanted to build a second smaller house. However, they were able to proceed using a suitable stone found on their property. The stone used in the second house, quarried on site, is paler in colour (Figures 3.14 and 3.15). Concerns about the comparative strength of locally sourced stone can be addressed by increasing the wall thickness, and spreading the support of lintels and roof structure over a generous bearing area.

Here is one more hint about looking for a suitable stone supply. While it is good to keep an eye out for all sorts of clues while driving around, it is not a good idea to get distracted from the road. If a building or a garden wall looks interesting, resist the temptation to have a good look while driving. Park the car and go for a walk instead.

There is a lot to look at, not just the rock face, but also any local quirks and tricks that may have been employed by some long-gone stoneworker. Has any stone

Figure 3.14: Fitting snugly up to the curved wall, the sliding door joinery was built with a lean, matching the batter of the wall, house near Bacchus Marsh, Victoria.

Figure 3.15: Straight and curved external walls add privacy to the bedrooms, near Bacchus Marsh.

Figure 3.16: End wall of barn, the walls have survived long after the roof was gone, near Hepburn Springs.

been laid on edge, and how is it weathering? Are there cracks in the stone, or only in the mortar? Does the stone have the same look as the rocks in the ground closer to home? And how did they do their corners, with the same stone or something different? Look for indicators like these to establish whether you are onto a good source of local stone (Figure 3.16).

Table 3.1: Localities of Australian building stone

GRANITIC STONE	Granite			Other	GRANITIC STONE	Granite			Other
Victoria	**Red**	**Grey**	**Green**		**NSW & ACT**	**Red**	**Grey**	**Green**	
Aura (Emerald)		o			Adelong		o		Gneiss
Beechworth		o			Albury	o	o	o	
Ballarat/GongGong				Diorite	Barren Jack	o			
Broadmeadows		o			Bathurst		o		
Cape Woolamai	o				Braidwood	o	o	o	
Ceres/Geelong				Diorite	Breadalbane		o		
Colquhoun	o				Bowral				Trachyte, syenite
Cudgewa				Porphyry	Bredbo		o		Porphyry
Dergholm/Casterton		o		Diorite	Broula Hills	o			
Dromana				Diorite	Bungendore	o	o	o	Gneiss
Everton		o			Burrowa		o		Porphyry
Gabo Is.	o				Canberra/Hall				Porphyry
Harcourt		o			Carrick	o			
Harrow		o			Cooma	o	o		Gneiss
Inglewood	o				Cowra		o		Porphyry
Korong	o				Goulburn/Pomeroy		o		Gneiss, porphyry, diorite
Kyneton		o			Grenfell	o			
Morang		o			Gunning/Collingwood		o		
Mt Taylor				Porphyry	Harden		o		
Orbost				Diorite	Inverell	o	o		
Shelley		o			Jerangle	o	o		Diorite
Talbot	o				Jindabyne	o			
Tallangatta		o		Porphyry	Lithgow	o			
Trawool				Porphyry	Maffra	o			
Tynong		o			Michelago	o			Porphyry
Violet Town		o			Montague Is.		o		
Wangaratta/Warby Ra	o				Moruya		o		
Walhalla				Diorite	Mudgee	o			
South Australia					Mulloon Ck	o			
Calca	o				Murrumbateman	o			Porphyry
Charleston Midgee	o				Oberon		o		
Gawler Ra				Porphyry	Orange/Canoblas Mt				Trachyte
Kingston			o		Rylstone	o			
Minnipa	o	o			Tarago	o			Diorite
Moody		o			Tarana	o			
Murray Bridge	o	o			Tenterfield		o		
West Is.		o			Trial Bay	o	o		
Wudinna	o				Tumut/Wee Jasper		o		Diorite
Western Australia					Uralla		o		
Gwambygine	o				Yass		o		Porphyry
Jerramungup	o				Young		o		
Karlgarin				Porphyry	Wombeyan	o			
Meckering		o			**Tasmania**				
Mundaring		o			Gladstone		o		
Namban				Porphyry	**Queensland**				
					Herberton				Porphyry

KEY
o single colour
+ mixed colour

NOTE: This reads like I've been everywhere, but few quarries are verified. Some places did not even yield local knowledge.

References:

R.T. Baker, *Building and Ornamental Stones of Australia*, Sydney, 1915.
Mines Department, *Victorian Building Stones*, Melbourne, 1949.
WA Dept of Mines, *A Guide to the Building and Facing Stones of Perth and Fremantle*, 1984.

Localities of Australian building stone – basalt, sedimentary, metamorphic													
BASALT	**New South Wales**		**Black|Grey|White**							**Grey|White|Red|Other**			
Victoria	Dundas	**MARBLE**	**Red|Multi|Other**						**Sth. Australia**	G	W	R	O
Footscray	Inverell/Sterling	**Victoria**	B	G	W	R	M	O	Angaston	+	0	+	
Lethbridge	Jerangle	Buchan	0	0	0				Kapunda	0			
Malmsbury	Kiama	Lilydale					0		Macclesfield	0	+	0	
Port Fairy	Orange	Limestone Ck		0		0			**W. Australia**				
SANDSTONE	**NSW & ACT**	Mansfield			0				Mt Wall	0			
Victoria	Albury	Mt Tambo	0						Nanutarra		0	+	+
Apollo Bay	Barbers Ck	Orbost (Martins Ck)			+	+	0		Wyloo		0	0	
Bacchus Marsh	Braidwood (Nerriga)	Thomson R	+		+		0		**Queensland**				
(Darley, Bald Hill)	Bundanoon	Toongabbie	+		+		0		Calliope		0		0
Bairnsdale Mt Lookout	Canberra	Waratah Bay	+		+		0		Chillagoe		0		0
Ballan	Galong	**New South Wales**							Gigoomgan Head				0
Barrabool Hills	Gosford	Abercrombie Ck			0			0	Herberton				0
Briagolong	Goulburn (Frogs Hole)	Adelong			0			0	Hunter Is.		0		0
Dunkeld (Mt Abrupt)	Marulan	Attunga				0			Iron Is.		0		
Egerton	Milparinka	Bathurst			0		0		Kangaroo Hills		0		
Greendale	Mundooran	Bibbenluke	0		0				Marblestone	0	0	0	
Lauriston	Newcastle	Binalong				+		0	Moreton Is.	0	0		
Longwarry	Ravensfield	Borenore				0		0	Mt Roundback		0		
Mansfield	Sydney/Hawkesbury	Brundle Ck			0				Rosedale		0		0
Stawell (Grampians)	(Pyrmont)	Buckeroo						0	Townsville		0		
Wangaratta	(Wondabyne)	Bumbaldry				0			**SLATE**				
Tasmania	(Waverley)	Bungendore		0					**Victoria**				
Hobart	Wagga (Grong Grong)	Burrowa			+	0			Castlemaine				
(Bellerive)	Yass	Caleula		+	0	+	0	+	Mansfield				
(Brighton)	**South Australia**	Caloola			0				**New South Wales**				
(Brynestyn)	Murray Bridge	Coolalie			0				Bungendore				
(Campania)	**Western Australia**	Cooma (Tollbar)			0				Chatsbury				
(Domain)	Donnybrook	Fernbrook					0		Cooma				
(Knocklofty)	**QUARTZITE**	Goulburn (Sheas Ck)				0			Grattai				
(Lindisfarne)	**NSW & ACT**	Havilah			0				Gundagai				
(Oakhampton)	Burrowa	Joir						0	Jerrawa				
(Risdon)	Queanbeyan	Kempsey						0	Queanbeyan				
(Spring Bay)	Tarago	Marulan					0		Taralga				
(Teatree)	Tarrabandra	Michelago (Norongo)					0		Towrang				
(Waterworks)	Uriarra	Molong		0					**South Australia**				
Patersonia	**Western Australia**	Mudgee						Gn	Mintaro				
Ross	Toodyay	Orange/Douglas	0		0				**SERPENTINE**				
LIMESTONE	**New South Wales**	Parkes/Ashburnham			0				**New South Wales**				
Victoria	Cowra	Portland				0			Gundagai				
Batesford	Dubbo	Queanbeyan					0		Lucknow				
Geelong (Galena Pt)	Taree	Rockley	0		0				Macleay R				
Moorabool	**Western Australia**	Rylstone	+	+	0	0		0	Nundle				
Mount Gambier	(Tamala limestone)	Springhill	0						Pt Macquarie				
Portland	Cottesloe	Tarago	0		0				Tarrabandra				
Sorrento	Fremantle (Rocky Pt)	Tumut		0					Tumut				
Warrnambool	Moore River	Warialda	+		0	+			Warialda				
Waurn Ponds	Rottnest Is.	Windellama	0						**Queensland**				
SOAPSTONE	**Western Australia**	Wombeyan			0				Canoona				
South Australia	Meekatharra	Yass	0						Kandanga				
Mt Fitton	Three Springs								Rockhampton				

Compiled from references on sheet 1, quarry trade information and on-line data

Figure 4.1: End result of quarrying: the remaining solid rock face and a pile of broken rocks. After being loosened up with explosives, a lot of excellent rocks were removed using hand tools.

'... laden with a cargo of spades, hammers, chisels, pickaxes and crowbars (we) began our undertaking. On the smooth face of the perpendicular rock I drew out in chalk the size of the proposed entrance, and then, with minds bent on success, we battered away. Six days of hard and incessant toil made but little impression; ... but we still did not despair, and were presently rewarded by coming to softer and more yielding substance; our work progressed, and our minds were relieved.'

Johann D. Wyss (1814), *Swiss Family Robinson* (English edition)

4

Small-scale quarrying

By quarrying your own stone, you have the advantage of first choice of the stone as it comes from its source, in the ground or at the quarry face. If you have stone brought in from a supplier, the same degree of selection is unlikely. Other people may have developed a quarry to sell stone to the public, but they do not necessarily have a better stone. If another quarry is close, and part of the same geological formation, chances are your stone is similar.

On the downside, quarrying is an extra job in preparation for building, and it adds time and effort to the construction process.

To get stone by quarrying, rough-shaped stones are separated from a larger mass of rock, sometimes referred to as the living rock. The source of the rock should be selected with a lot of forethought. If it is beautiful, then perhaps it is best left alone. If it is part of a cutting along a driveway, or if it has been quarried before, then it is worth exploring. Even better if the track that leads to your house site turns up good stones in the course of road-making. Living rock is a part of nature. It may be covered with lichen and moss. Lichen can take decades, or even centuries to colonise a rock surface.

Once removed from the quarried rock mass, stones are broken into sizes that can be handled more easily. The ideal sized stones are the largest that can be lifted into place by one person as a wall is built, as this helps it to go up quicker than a lot of little stones. It is good if the stones have a high vertical face as well as a generous bedding surface. With smaller stones, more time is spent fitting them together, and more mortar is used. Of course the small stones are still necessary in between the big ones, but usually there are enough smaller stones to choose from, and bigger stones can always be broken into smaller stones. (This sometimes happens when you don't want it to.)

Stones of the preferred shape and handling size will be in greater demand at the building site, so a rock formation that yields these consistently is the best place to quarry.

Some stones are too large to move easily, and with no clear lines to indicate cracks. The odd one of these may still be good to use intact as a lintel, a shelf or, if they are too heavy to easily manoeuvre, somewhere low in the building, such as a solid step. Such details may require a slight change in the design of the building. If large stones are left exposed to the weather for a while, they may show up a crack in future. If the stone is grainy and workable enough, it is possible to split it into two using drilling and wedges. If it is brittle, attempts to split a stone may end up shattering it into awkwardly shaped pieces.

Quarrying may be done with hand tools or power tools, and in some cases explosives are used (although care and expertise are necessary to avoid a high proportion of shattered stone). For a house, the amount of stone required is relatively small, and it should be cheaper to use hand tools. For large masses of solid rock, it may be desirable to drill holes as well, to start crack lines in the rock.

Sedimentary stones and slates have layers, so they will split more easily in one direction. The layers of sedimentary stones can also deteriorate and peel off when exposed to weather if laid face on edge. This depends on the stone and its location in a building. It is therefore important, with cut stone of this type, to know the direction of the sedimentary layers so they can be laid properly. If it is hard to tell once the stone is out of the quarry, it is a fair assumption that the stone warrants being marked for the direction of layering.

With recycled stone, such as from a church spire that has been replaced, it may take close inspection to recognise the layers of sedimentary stone. The way it was placed and where it came from can verify this. Laying the stone the same way helps it survive weathering. Where walls are not exposed, say sheltered by a roof, this is less critical.

Machine quarrying

To quarry large quantities of rock, in a commercial quarry, it is economically viable to use machinery. This is expert work, and should be picked up from the trade, not book-learning, so this description is a rough guide only, for those buying or specifying. For housing, the expense of quarrying with a machine is usually prohibitive, although it may be viable for paving and facing stones.

Marble is usually cut into blocks of the required size by channelling machines. These cut out a narrow channel, thus reducing waste and avoiding the formation of cracks. Granite may be quarried in the same way for ornamental work, but it is more often quarried in other ways, as wastage is not as much of a concern. In a large outcrop with few natural cracks, the rock is drilled. Then a section of rock is

Figure 4.2: Working a line in a stone to open a crack with the blockbuster, quarry on property for house near Hepburn Springs, Victoria.

separated from the quarry face by 'broaching', which opens up the drill holes, forcing a crack.

Large-scale operations involve the use of cranes to move heavy blocks of stone from the quarry to the saw bench and cutting machinery.

In commercial quarries, if marking of the layer direction is needed for sedimentary stone, chances are it is being done. In smaller operations, the rock is broken up in the quarry until it is of a reasonable size to be handled and transported away. There may be various shapes and sizes.

In cases where the stone has been quarried using explosives, lines of weakness may develop. If the stone is then cut, these lines of weakness do not crack open as readily as they would if worked with a chisel. If the stone is used in a thin veneer, problems may occur later due to cracking when movement occurs.[1]

Quarrying with hand tools

Hand quarrying is most feasible where the rock formation or quarry face has plenty of joints or straight cracks (Figure 4.2), so smaller stones can be separated

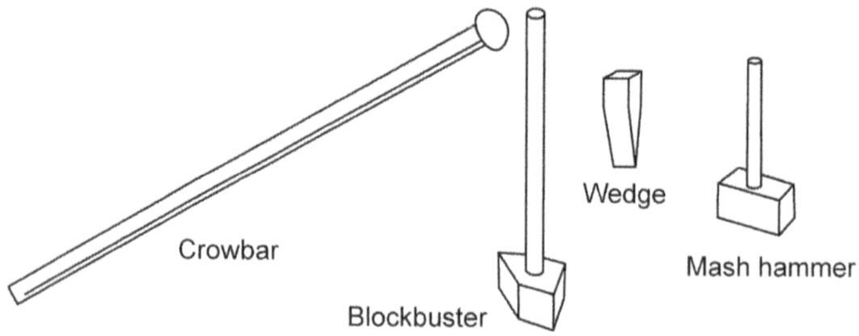

Figure 4.3: Hand tools.

from the rock mass without excessive effort. While not an easy job, it is rewarding to collect a load of fresh rock, even if it has to come with a few blisters at first. The most readily available tools for hand quarrying are the crowbar, blockbuster (or sledgehammer), wedges, and of course a pick and shovel (Figures 4.3 to 4.5).

A rock formation may have stones of a good size and shape for building at, or just below, ground level. These can be prised from the exposed surface by leverage with the crowbar. Sometimes it is necessary to clear the way first with a pick and shovel to get a better hold. The rock is often consistent in quality for some distance each way, and once the overburden is removed, it should be possible to follow the joint lines and lift more stones with the crowbar and wedges.

Cracks may need to be opened up with the blockbuster to make the rocks break into suitable sizes. Where the rocks are tightly packed together, they may be difficult to pull out of the quarry face. By moving the rocks back and forth with the crowbar, and opening up cracks to fit the wedge in with a blockbuster, the rock edges can gradually be loosened from each other.

Figure 4.4: Levelling off the wall for the ground floor after quarrying out the cellar, house near Hepburn Springs.

Figure 4.5: Looking into the cellar while work was under way paving the floor. Most of these rocks were quarried on the spot. Note sump at lowest point, draining downhill.

If the rock needs a bit more persuasion, it helps to have two people doing this work. As the rock is levered more and more from the quarry face, the cracks get wider. As stones are prised out they can be held from slipping back by using smaller rocks, or by leverage with wooden poles. A wedge can be driven in at a crack, using the blockbuster, and care should be taken that it is not positioned where it might tighten up the cracks and make it harder to remove the rock. A mash hammer can also be useful to put the wedge into cracks (where less force is required).

Wooden lever poles are useful in lifting heavy rocks, as the large weight of rock concentrated on a steel crowbar may cause it to bend. The crowbar can be used first to open up tight cracks, and then the timber poles can be fitted in where there is space to get leverage.

More challenging rock may call for some of the expertise of the stonemason's trade. Large blocks of stone with no cracks or lines of cleavage can be quarried with specialised tools. These tools are used in quarries or by stonemasons, but are not widely available. Some stonemasons make their own steel tools. It may be difficult to get advice on where to buy suitable tools, but local knowledge is a good start. If it is not currently happening in the local area, it may be worth seeking out a stonemason in another location, where stonework is more common.

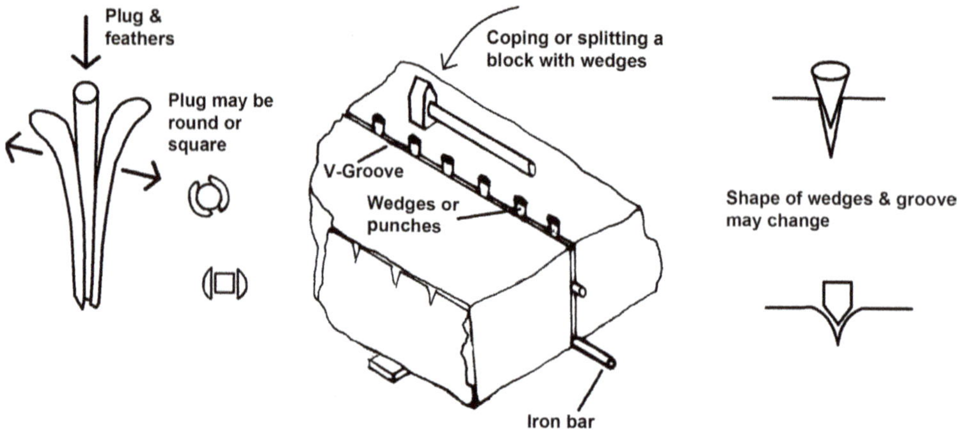

Figure 4.6: Traditional tools.

The plug and feathers were traditionally used where holes had been drilled first, a few inches apart (say every 100 mm). These exert a great pressure and force a crack in the direction in which the feathers are aligned. The feathers are lined up to push outwards from the drill hole in the direction that the stone is to be split. This method is used for splitting a stone into two similar-sized blocks (Figure 4.6).

With suitable stone, a block that is too large can be split by 'coping'. First a V-cut is chased in on top and along the sides where it is to be broken. A wooden 'skid' is put under the stone together with an iron bar along the required line of splitting. Another skid is placed to sit the rock on, and a short block placed at the end just below the rock for it to fall on when split. Wedge holes are made along the V-cut with a sharp chisel (punch) or masonry drill. The bottom tip of the wedge should not rest on the bottom of the hole or it will jump out when hit. Once the holes are cut, the wedges are inserted in each one, carefully so they point in the required splitting direction. They are gently tapped at first with a heavy hammer until they have a hold, then harder blows are made on the wedges with equal force in quick succession until the rock splits. An old-style Japanese technique is similar, using a curved V-incision every 10 cm and inserting a small wedge (mild steel, about 3 cm wide) in each one. These are then hit in succession with a sledgehammer.

Quarrying with explosives

The use of explosives for quarrying can be helpful in some cases, but it also may create extra waste, as the rock can shatter into irregular pieces (Figure 4.7). There is a chance of extra cracking in rocks quarried this way. Stone within a metre of the blast can lose strength and durability. If the stones are worked with a blockbuster

Figure 4.7: Quarrying with explosives is not always successful, as the quality still depends on the rock formation. If natural cracks do not help rocks to break neatly, they can be too large to use.

and chisels, they will crack where the weakness occurs. If a rock is sawn without regard to its internal lines of weakness, it is liable to crack when subjected to loads in the building.

A licence is needed to use explosives. If you are quarrying rock and you want to use explosives, it is best to hire a blasting contractor, who should have knowledge and experience of working with rock, as well as providing the necessary drilling equipment.

Typically, before blasting, one or more holes are drilled to the required depth behind the desired rock or rocks in the quarry face, and then small charges of coarse black powder are put into the holes and fired together. Several repetitions of lightly tamped charges should work better than the same amount of powder in one blast. For a single block of stone, one shot-hole may be enough to blow it off cleanly. Larger rocks require two or more holes.

Black powder is recommended to loosen a layer of stone from the bedrock. Apparently dynamite is not as effective, as a heavy explosion is more likely to shatter the rock and form unwanted cracks in the good rock. Some experience is necessary to know the best place to drill the shot-hole, as wrong positioning can ruin the stone.

More success is likely if blasting is done where the rock layer is located in such a way that it can 'peel' away from the adjacent bedrock than if the charge is placed in a quarry pit that does not allow sideways movement of the stone.

Figure 4.8: Small granite quarry for house near Moruya, New South Wales.

Granite

For granite, the quality of the rock is better deeper down, where the rate of cooling was slower. The deeper rock is more homogeneous and has larger and more decorative feldspars. The joints in granite rock are usually in two directions, horizontal (or parallel to the surface) and vertical, with occasionally a third set running diagonally (Figures 4.8 and 4.9). The intervals between joints can vary considerably with different rock formations.

Where the joints in granite are wide apart, they can be opened up using wooden wedges. These are driven in dry, and water is then added to expand the wedges and open up the gap (this method was used by the ancient Egyptians). Where the joint cracks are closer, the stones can be broken apart with the crowbar, blockbuster and metal wedges (Figure 4.16).

Large blocks of granite without cracks used to be broken up by drilling holes and driving in plugs and feathers. This method is not in common use now.

Basalt

Basalt can be hand quarried where it occurs as a columnar formation. In this case the lava has formed prismatic columns as it cooled. These prisms, which may be roughly hexagonal, can be broken up into building stones of a workable size.

Figure 4.9: Granite walls of house near Moruya, New South Wales.

Sometimes the rocks are broken up already by the weathering action of water or by tree roots.

The surface rocks of a basalt lava flow are usually quite porous and full of air bubbles. They are unlikely to have a flat top and bottom for building into a wall, but they can be built into a rough rubble wall. In areas with a lot of this stone, it is common to find dry stone walls built with them, which was done to clear the fields so they could be ploughed.

Although it may show up an occasional air bubble, basalt has few impurities. It is fine-grained but very hard, so it is time-consuming to shape it. However, it can be worked to regular blocks if wanted, and it is very durable.

Sandstone and other sedimentary stone

For sedimentary stone, the quarrying technique used will depend partly on how homogeneous the stone is, and the thickness and angle of the bedding layers, as well as on the frequency of lines of weakness or joint lines in the stone. Joint planes may occur in two directions approximately perpendicular to the bedding planes. If they are at right angles the rock can be broken into ideal stones for building, but they are often aligned at some other angle. The existence of joints makes quarrying easier, and also puts a smooth surface on the side of a stone, which is ideal to use as the face when it is laid in the wall. Often the joint face is a slightly different colour from other roughly broken stone. This is due to the presence of soluble salts that have been deposited on the jointing surface, where movement of moisture is greatest in the rock. The location of such joints can be determined by lines of a different colour on the surface of the unbroken rock.

Narrow cracks are often filled with a thin layer of clay and sometimes roots. These can be fairly easy to break open with a blockbuster and crowbar, but they may not penetrate across the width of the rock, so the crack may not break into a clean surface all the way across it. By working the opposite side of the stone, it is possible to put a new break in the stone roughly in line with the crack. Hopefully these will align to form clean edges for the split stone.

Once the natural cracks are opened up, it should be possible to move each rock in turn by leverage with a crowbar and wooden levers. If the bedding plane has been tipped to a near-vertical angle, the rock layer can be quarried downwards from the surface along the joint lines, by dislodging each rock with a wedge and blockbuster, or it can be quarried from the side by removal of the overburden (Figures 4.10 and 4.11).

When the bedding plane is horizontal or parallel to the surface, the overburden must be moved to a position where it will not impede future quarrying. When the bedding layer is exposed as a quarry face, it should be possible to lever the stones out of it, one at a time, from the top downwards. If a stone cannot easily be disengaged from the other rocks in the bedding layer, it may be necessary to loosen a larger area of the rock layer from the bed alongside. When the adjacent rocks are pushed

Figure 4.10: Rocks of a good size at a quarry close to the site, house near Hepburn Springs. There is a pile of overburden from the quarry on the left, and the pickaxe is being used to open up a further supply of rock along a seam. This quarry yielded a lot of rock before being levelled for the garden.

Figure 4.11: Prying out rocks with a crowbar for loading into the ute.

outwards from the quarry face, it should be easier to pull or lever one away from the others, which frees the rest of the stones in the bed. The ease with which this operation can be done depends on the angle of the bedding plane, and the amount of overburden holding down the rocks around the sides.

The colour of the rock surface can change after quarrying, as the moisture moves to the surface and evaporates. Iron oxides are the most common colouring agent, and give a red or yellow colour to the stone. The colour that develops after the separation of the stone from the bedrock is the result of the drying of dissolved minerals in the stone. Once the stone is dry, smaller stones taken from a rock mass may not exhibit the same characteristics. If a very large block is extracted and allowed to dry, there can be no assurance that stones cut from the inside will match the outside colour.[3]

Other mineral solutions can act as a cementing material, but are water soluble. Stones with carbonate of lime holding them together are vulnerable to weathering in exposed locations.

Silica or silicate deposits may occur between the surfaces. These have formed with a lowering of temperature, or chemical change, and can appear as regular bands in the rock. They may be in the form of quartz crystals. Rocks cemented with silica are durable. The rock may be easy to split along a quartz vein, but often the silica is well bonded to the older rock.

Flagstone splits easily because it has layers of mica within it. It will form good blocks of stone with a slight sheen along the micaceous surface. Workability in other directions depends on the composition of the sediment that formed the other layers. Flagstone is commonly known to be suitable for paving stones because of how cleanly it can be split (Figure 4.12).

Figure 4.12: Flagstone quarry near Bairnsdale, Victoria.

Slate

Slate is a harder material than sandstone, and has more distinct layering. This makes it possible to split into thin pieces, in parallel planes. It is of variable quality. Most slate quarries yield a stone that breaks into thin slabs of irregular shape.

Very rarely, a seam of slate is uncovered that is hard enough and splits thin enough to be used for roofing slates. Commercial slate quarries generally use diamond-tipped saws to cut the slate into regular tiles.

Slate will split easily along the layers of cleavage, but it is difficult to break cleanly in the other direction unless there are jointing planes. Sometimes a layer of a slightly different colour is visible along the edge of the stone. This line can be opened up with a broad chisel to split the stone. The colour of edges like this is more varied than the rest of the stone, and may show up reds, yellows and other colours not found in sawn faces. The broken surface shows up impurities that were formed in the stone by the movement of moisture along that layer.

Usually slate is used for flooring or sills. In some places, locally quarried slate may be suitable as a damp-proof course, but it would be risky to use without a waterproof mortar additive. Second-hand roofing slates may be available instead, and can be used for damp-proofing, in two courses with overlap.

Limestone

Limestone is a softer material to work than other stones, and at its source it has not been contorted, weathered or buried to the same extent as older rocks. A pure limestone is white or pale grey. The majority of grains in the stone are broken-down fragments of shell or sand particles, cemented by calcite or dolomite.

The stone may be veined or mottled with remnants of shells, corals and marine creatures of varying colours, such as red, pink or black. Some limestones contain

Figure 4.13: Very large rocks, useful for footings or for lintels if not to be lifted too high.

ooliths, a spherical build-up of calcite deposits around particles (this stone is also called oolitic limestone). Compact limestone breaks with a smooth surface, and not a very granular structure. It can be polished and is sometimes used like marble.

In some instances, limestone that is soft when freshly quarried will harden on the surface when 'seasoned' by weathering.

A limestone formation generally lies in well-defined, near-horizontal beds, which vary in thickness up to a maximum of 2 metres. It is possible sometimes to quarry limestone entirely along natural beds and joints. It can be found close to the surface. Some limestones are quarried out in dry horizontal galleries below a seasoned layer at the ground surface.

Limestone may often be soft enough to be hand sawn. A specialised saw can be set up for cutting blocks of stone. It should have water running over as the stone is sawn. A drip board may be used, where water trickles down a slope with sand that is gradually carried into the saw-cut with it (this procedure is described by Purchase (1896)).

Loading and unloading

At the quarry, the good building stones are piled up in a place accessible to vehicle access, while the odd-shaped excess rocks are placed alongside the overburden so as not to get in the way of future quarrying activity. Stones that are marginal can

Figure 4.14: Large lintel rock being loaded over cellar wall by leverage, onto 44-gallon drums, planks and wedges, before being settled onto a deep bed of mortar each side.

be put nearby, to be sorted later on. Some may need a bit of chisel work, or may be just right for an odd-shaped section of wall at a later stage.

The truck or utility should not be overloaded. Stones are best placed as far forward as possible in the tray, to distribute the load between the wheels. Otherwise, regular use can put undue stress on the back springs and tyres.

A station wagon can also be used, but it is difficult to keep the inside of the car clean, not to mention free of creepy crawlies. A sheet of plywood can be used to avoid damaging the floor surface when rocks are thrown in.

The building site must also be organised to receive the load.

The variety of stone that can be collected is immense. As the stones are fitted on the wall, there is constantly a need for a stone just 'this wide' or 'so high' to fit alongside the others. The goal of finding the right stone to satisfy a picture in the mind's eye can send a stoneworker walking purposefully around pile after pile of stone until the right one is recognised.

If the stones are organised according to size and shape as they come off the back of the truck, the searching process is much simplified.

The biggest stones will need to be dropped right next to the tailgate, or in some cases, directly onto the wall (Figures 4.13–4.15).

Figure 4.15: The large lintel stone for the cellar window goes the full width of the wall.

The smaller ones can be chucked further away, and organised in segments around the unloading area according to whether they are blocky, uneven, long, flat or wedge-shaped.

Potential corner stones, sills, paving stones or arch stones will then accumulate. When there is a need to look for more of a particular kind of stone in the quarry, it will be evident because the particular section of the stockpile is getting empty.

Figure 4.16: Granite quarry for house near Moruya.

Figure 5.1: Arched window in rock-faced wall, curved voussoirs and a smooth finish to the window reveal.

'The stone may show a natural face in the wall, or a face characteristic of the tool used to shape it.'

Frank Lloyd Wright (1928), *In the Realm of Ideas*

5

Shaping stone

The amount of work done to shape the stones will greatly affect the time spent building a wall. Even the roughest stones can be fitted together with plenty of mortar in between as a filler, to get a wall built quickly. It is the time and effort taken to fit stones together closely that can make stonework a slow process. The more shaping that is done to the stones in a wall, the closer they will sit together. This reduces the amount of mortar needed, which is an advantage, but adds to the number of stones used and to the time spent building the walls of a house. Another factor is the size of the stones. The best size is the largest that can easily be lifted to the height of their position in the wall, by one person (bending the knees, not the back).

Dry rubble walls require more care in fitting the stones together than walls built with cement mortar.

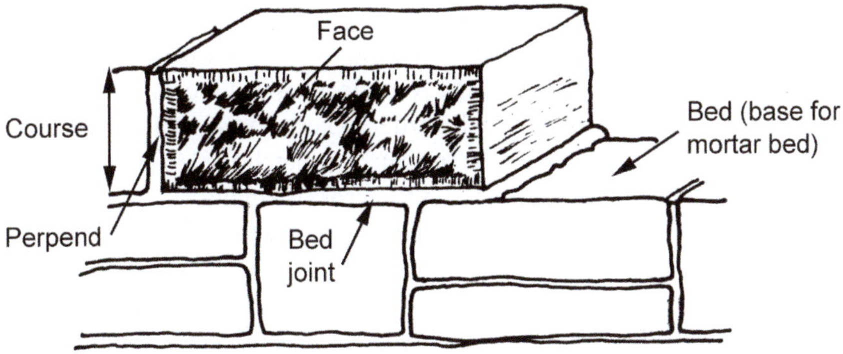

Figure 5.2: Building surfaces (i.e. in contact with mortar or adjacent stone).

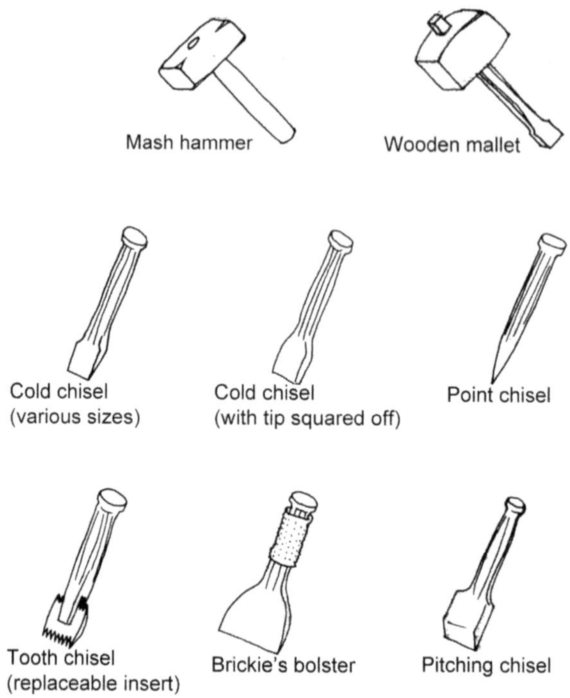

Figure 5.3: Hammers and chisels.

Ashlar is the most time-consuming type of wall, as all the building surfaces of each stone (with adjacent stones) are worked to a smooth, close fitting finish, and its face surface is usually worked too. Ashlar or wrought stone was more common before reliable cement became available. Some old stonework is joined with joggles, or hidden connections, so there is just grout between stones.

Rubble stonework is more usual for houses. The amount of shaping depends on the pattern of laying the stones in the wall (Figure 5.2). A flat base and top to each stone makes the wall stronger than if the stones are rounded, as rounded or rougher stones rely more on mortar to stop them settling unevenly. It is unlikely that rubble stones will all be ready to fit into the wall in their natural state. Rough stones do not fit together closely and the result is oversized mortar joints. Bumps on the top or bottom of a stone make it harder to keep a level surface for the next course. Sometimes a stone needs to be trimmed to match the level of the one beside it, so that a stone in the next course can span across them both.

Shaping the face surface of a stone is only necessary to get rid of rough or jagged protrusions, and even that is a matter of preference. It is hard to avoid chisel marks on a worked stone, and the natural rock face may be spoilt by unnecessary chisel work. For a worked rock face, it is better to chisel back from the centre of the face, progressively working towards the outside to break off chips from where the

Figure 5.4: Trimming a stone to fit, working up on a scaffold next to the wall being built.

stone already shows chisel marks. In ashlar work the stone is often rubbed smooth to avoid chisel marks, but this creates a different texture from the natural rock face.

If there are a variety of stones with rough and smooth faces, the smooth stones can be selected for interior stonework, and the rough stones used outside, or higher up in the wall.

Achieving consistency in the range of stone sizes and joint thickness is desirable. It is especially important to maintain a consistent standard where more than one person works on the walls. Some stones are brittle and difficult to work, so only a minimum amount of trimming should be done, and even then there is a possibility that the stone will break up if it is hit too hard in the wrong place. A hammer and stone chisel or point are suitable for rough shaping of most stones as they are laid (Figure 5.4), and for some stones a smoother finish is possible.

The weight of the hammer used is a matter of preference, but is usually related to the hardness of the stone. A metal head such as that on a mash hammer is ideal for harder stone and is a good size for chisel work (Figure 5.3). For softer stone such as sandstone or limestone, a wooden mallet is better, as it causes less fatigue or jarring of the hand from constantly working the stone. Rubber mallets are sometimes used.

Unfortunately for the hands, a steel mash hammer is better to provide the necessary force on the chisel for harder stone such as granite.

A few different chisels are useful, each being preferred for a certain type of work. Ideally, chisels should be of hard-tempered steel or have a tungsten tip. An ordinary stone chisel should be adequate for most trimming work with softer stone. A point chisel is less likely to disturb potential cracks within the stone and is useful for removing irregularities on the surface. The point may be preferred for harder stone. A wider chisel such as a brickie's bolster or a pitching chisel is good for making a straight edge along the stone where it strikes. The tip of the chisel may work best if squared off to form a flat edge rather than a point. The advantage of this is that the weight of the chisel pushes more of the excess stone away as well as making the edge line as it hits. Toothed chisels can be used in some areas with a good local stone. They can smooth off the surface, but leave obvious marks that are not so good on the face.

A stone coming from near the top surface of a rock formation will have been exposed to weathering for many years. The constant percolation of water opens up the natural lines of weakness. In some cases this gives each stone a smooth face where it meets another. Calcium carbonate in the stones may have dissolved and recrystallised on the surface as calcite to cement the particles together. These surface stones are harder and more brittle than stones cut from deeper in the rock formation. As they are not easily worked, except to knock off rough edges, they are best selected for their shape when collected, and not shaped much more afterwards. Weathered stone such as field stone may be brittle, and can break in unexpected directions when hit with a chisel. Cracks are visible in some stones. If cracking appears to penetrate deep into the stone, or contains roots, soil or organic matter, it shows a line of weakness in the stone. Often a stone with a surface crack is strong enough to build with, but if it is to be laid in an exposed part of the building, where it might deteriorate further at a structural loading point, it is wise to split the stone into two smaller but stronger stones.

Where lines of weakness are visible on the surface of a stone, cracks can be opened up with the point of a blockbuster to good effect. If there are a number of lines of less prominence on the surface, it is worthwhile hitting the stone with the hammer end of the blockbuster first, to see if the impact will start a crack along whatever line will split most readily.

For trimming rough stone to fit better, a hit with the mash hammer may be adequate. If the stone is brittle, only a small amount should be removed at a time. A whack with the edge of the mash hammer at the required place may be enough to knock off a small chip of stone. For breaking off larger bits of stone, a chisel or a brickie's bolster is better (Figure 5.6).

A brittle, layered stone is difficult to shape (other than to split between layers) and it is not worth spending too much time on it, except to trim small amounts off

Figure 5.5: Rock-faced wall with protruding stones, worked arris to corners, Castlecrag, New South Wales.

so that it fits into the wall. This kind of stone may shatter if hit too hard, or break into an unsuspected plane of cleavage. To trim off a rough bit of any size, only successive small bits should be tackled.

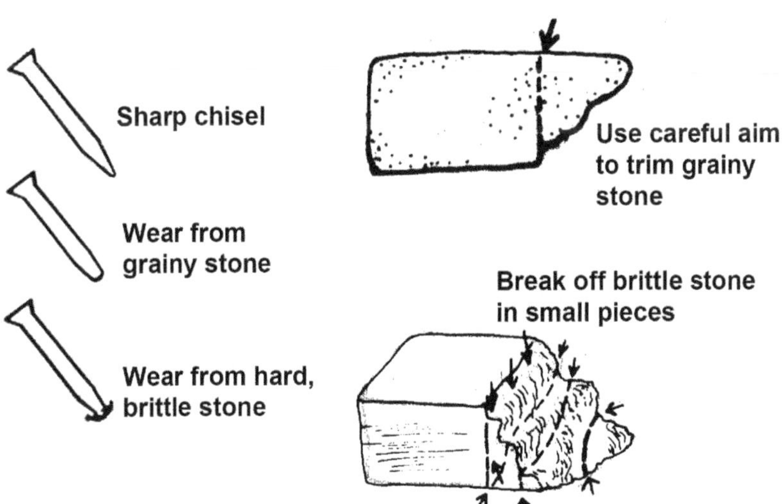

Figure 5.6: Trimming rubble stone.

With a good sandstone or limestone, the stone has a grainy texture and it is possible to control the direction that it will chip away with the chisel. It is important to maintain the shape of the end of the chisel with a sharpening stone. If the stone being worked is fairly soft and grainy, the flat tip of the chisel will erode down to a rounded end, whereas with a hard, brittle stone, the tip will flatten further with the constant impact, and the metal at the point tends to bulge and spall away around the sides.

For harder stones with a grainy texture, such as granite, the chisel can be aimed further into the block of stone to break pieces off. The chisel should be directed at an angle that compensates for the resistance of the stone, to get the desired splitting angle. The split in the stone tends to follow the shortest path across the stone from where the chisel hits, closest to the angle of the hit. Finding the right chisel angle takes a certain amount of trial and error, but this is part of getting to know the stone. It may be necessary to waste a few stones to get a better idea of the characteristics of a particular type of stone. It takes a while to learn how to send a shock wave through the stone so it splits in the right place.

To trim the surface of a stone that is already an easy size to handle, it may be better to chip off small amounts at a time. With chipping, as with splitting stone, it may help to aim deeper into the stone, as the excess stone chips off at a shallower angle than how the chisel is aimed.

On the face, it is best to work from the centre outwards, choosing carefully where to strike the stone to avoid obvious chisel marks, and ideally finishing by working in from the edges to chip off the previous chisel marks.

The top and bottom of the stone generally require more work than the other surfaces, as they determine the way the stone sits on the wall and its height in relation to the course or the stones on either side. The sides of the stone need only be worked to give a neater fit with the stones alongside, and in most instances the back of the stone needs no work at all. Working the sides or the face of the stone may be done for aesthetic reasons (Figure 5.7), to reduce joint thicknesses and to smooth off the wall surface.

Sedimentary stone has a distinctly layered structure. The bedding layers should be laid horizontally in the wall as this direction is strongest and has the most resistance to weathering (Figures 5.8 and 5.9). If sedimentary stones are to be used for an arch, the layers should be placed in such a way that they radiate outwards from the centre of the arch. In some stones it is difficult to determine the bedding layers, and ideally they should be marked to show the bedding plane as they are taken out of the quarry.

Ashlar is not usual in house construction, and where used it is most likely to be for a few special features, and for doorways, windows or corners rather than for straight runs in the wall (Figures 5.8 and 5.10). Since wrought stone takes a greater degree of skill, this is an area where a qualified stonemason is best employed for

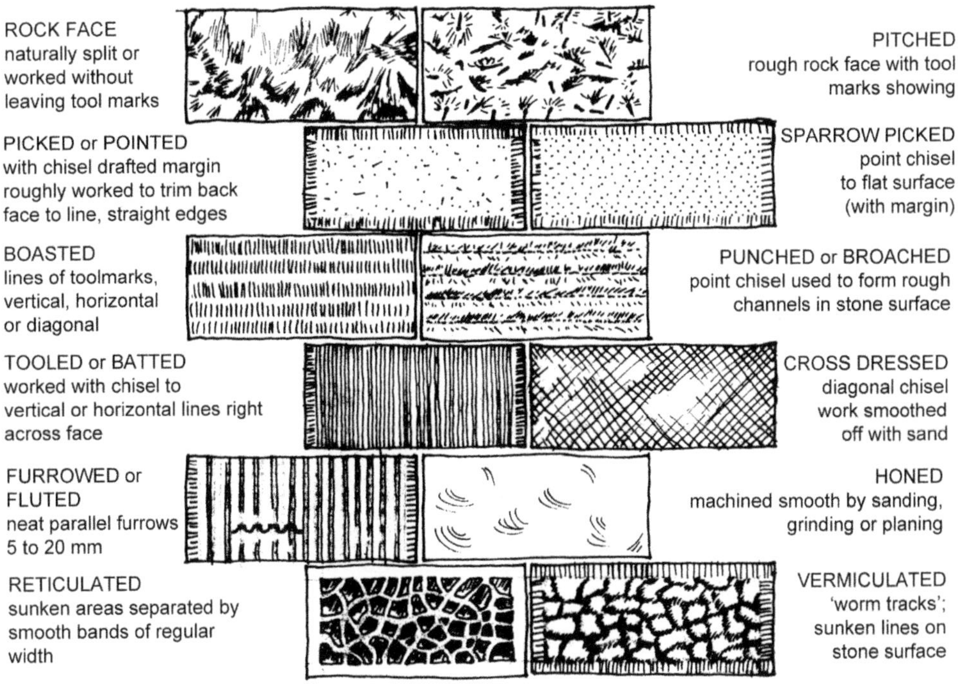

Figure 5.7: Finished faces of worked stone.

Figure 5.8: Weathered stones at base of a worked stone wall – pitched stone with margins to edges of quoins and lintels, Dubbo, New South Wales.

Figure 5.9: Weathered stone showing horizontal layers and spalling where a stone was laid on edge.

Figure 5.10: Unusual treatment of a window where stones each side are 'framed' rather than stepped for each course, Dubbo, New South Wales.

Figure 5.11: Arch cut in two stones (effectively a corbelled lintel) and stepped lintels, Hunters Hill, New South Wales.

the amount of wrought stonework required, and the majority of the walling, which takes less expertise, can be built by the owner builder.

For ashlar work, in addition to squaring the stones, projections or grooved recesses may be used for jointing. Such joints strengthen the stonework, and were once necessary, before reliable cement was available to bind stones. The precision work required for such joints is only feasible with stone that is easily worked, and it is most common in decorative stonework that is carved to some extent as well. Slender, free-standing stones are the most likely to require dowels or joggles, sometimes in addition to mortar between the stones (Figure 5.11).

In very detailed work, stones are shaped to fit together like timber joinery. A course joggle (like a tenon) protrudes from one stone and fits into a recess (or mortise) in another stone. It is used in such places as stone stair landings.

The joggle connection may be made from rectangular strips of a harder stone. Joggles made from slate or granite are set along the stone course, with half the joggle in each of the adjacent stones, forming a key to prevent lateral movement. Simpler joggles can be made with cement mortar as a filling in rounded holes, or shaped channels, cut in the stones where they butt against each other.

Figure 5.12: Hierarchy of lintel style, from front to back of historic gatehouse, Bringelly, New South Wales. The decorative moulding visible from the front is discontinued at the angled wall, and the back window has a smaller stone for the lintel.

Dowels are set deeper into each stone than joggles and are especially suited where a slender stone section such as a column or window mullion must be connected to the stone below. Stone dowels are generally square in cross-section, sometimes tapered to a smaller width deeper in the stone. Pipe dowels are now more commonly used, set into drilled holes.

Stone or metal cramps prevent stones from being pulled apart by lateral forces on the wall. If metal, they have a bend at each end that acts as a hook built into a notch in the stones. A stone cramp is shaped so it is narrowest where the stone blocks meet and is wider deeper into the stone. Ferrous metals (iron) should not be used because exposure to moisture can cause rust, which can start cracks from internal pressure due to an increase of size in the cramp.

Before attempting to do a lot of stone shaping, it is worth buying the right chisels. These may be hard to find through generalist hardware shops. Stone tool suppliers can be found, but usually only in districts where the local stone is easily worked and commonly used. Some blacksmiths may be familiar with the tools necessary for stoneworking. There are also stonemasons who make their own tools and might be able to supply them to others, but this is a slim chance.

We all admire a building with neatly fitted stonework and a rock face (Figure 5.14). The more natural the face, the more it brings to us the feeling of a natural setting, which makes us feel comfortable. In some stones, it is the way they break up along cracks that have been formed naturally that gives them their rock face. In stone that is worked, a rock face simulates stone that has been naturally chipped

Figure 5.13: Detail of lintel with decorative drip mould rising over the window, Bringelly.

Figure 5.14: Round window with crystal-shaped accents in rock-faced wall with log cabin style corners, Castlecrag, New South Wales.

away. In nature, the chipped edges usually tend to crumble or erode under the action of water. In building, our goal is to achieve an equivalent aesthetic effect (Figure 5.12), while preventing the stonework from suffering too badly from the ravages of nature.

Certain conventions have arisen as a result of the need to minimise damage from weathering. In a sedimentary rock, the direction matters. We use stones sitting flat in their natural bedding layers for the normal courses in walls.

For corbels and parapet tops, the sedimentary layers can be placed as books on a shelf. For arches, the keystone can also be placed this way, while in the voussoirs the layers radiate around, tilting a little more for each one. Slates are more likely to be impervious in either direction, but there is a lot of variation in stones that are called slate, sometimes it has some of the metamorphic characteristics but is little more than a hardened mudstone.

In historic buildings, a drip mould was commonly included to avoid having water run down a wall (Figure 5.13), especially in any exposed details of an upper storey.

It can be observed with stones from some quarries that they show less weathering when rock-faced than they would if cut smooth. When a smooth surface breaks up, it is obvious. This also depends on how well the stone is sheltered from the rain.

Figure 6.1: House near Young, New South Wales, squared granite corners, brick flat arch, and tuck pointing.

'Stone should be set in a wall on its natural bed as it was in the quarry, with the laminations horizontal, otherwise it will have a tendency to flake off the surface and to crack vertically. The exception is in a cornice, where the top members would flake off along the natural bed and in this case the stone is set with the bed vertical and at right angles to the front of the cornice.'

C. Lloyd (1958), *Building Construction for Craftsmen and Builders*

6

Techniques of wall building

Stonework is very versatile, having the capacity to be shaped and laid in regular squared-off blocks or built with randomly shaped and fitted pieces. It may be used as a veneer skin with a timber-framed wall, as structural masonry supporting two or more storeys, as a facing on other masonry, or it may be laid between formwork with concrete infill behind it.

It is generally used with cement mortar but can be built without mortar or with mud or clay mortar (Figure 6.2). The stones should be carefully placed for structural stability, on a firm support and with a good fit, before they are bedded in mortar. Mortar fills the gaps, and should not be relied upon for strength, which is the result of properly laid stones.

A stone wall imposes a heavy load on the ground below, so the first thing to do when building in stone is to ensure that the footings are adequate to distribute that load over the foundation material.

Footings

The footing is where the wall meets the solid ground or foundation. On a rocky site the bedrock is close to the surface and, if the rock is solid and difficult to break up, a reinforced concrete strip footing can be built on that, even where it is a bit shallower than the normally required footing depth. Old stone buildings were often built on stone or brick footings (refer to Appendix 1). On unstable foundations or on a flat site it may be cost-effective to use a concrete floor slab. In many cases, large rocks are located within a softer subsoil, as floaters. Under loading conditions, the differential movement of the material above these rocks compared with the other subsoil can be a reason for cracking in a wall.

Figure 6.2: Stone with mud mortar at Wilpena Pound, South Australia.

Stone walls need to be thicker than the usual single skin of masonry in a brick veneer wall, and this affects the loading conditions. Footings or slab edges should be wider than the intended wall thickness. Unless bearing directly on bedrock, the base of footings must always be level. On a sloping site, stepped footings are necessary, and must meet the required minimum dimensions. Attention should be given to the likelihood of heavier loading due to the locations of structural roof members or the placement of openings in the wall, to avoid cracking in the stonework when settlement takes place (Figures 6.3 and 6.4). To accommodate large stones at corners, the footings can be made a bit wider. Internal stone walls also need adequate sized footing beams below them.

The footings are a very important structural part of the wall. Where cracking occurs in a masonry wall it is often due to failure of the footings to withstand external pressures.

Reinforced concrete footings should be designed to comply with the relevant Australian Standards, to suit the soil type. For unknown situations, a margin of safety should be allowed. To avoid spending more on greater quantities of material, it is preferable for a structural engineer to design the footings based on the actual conditions.

Differential loading – doors and windows

The presence of tree roots or expansive clay and water can cause differential movement of foundation soil around the house, which puts stress on the footings. It is especially important around doors and windows to make sure that the footing

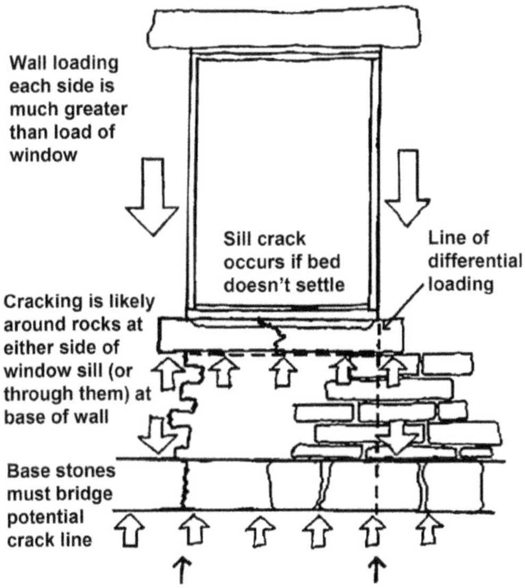

Figure 6.3: Differential loading and cracking.

and wall stones span across from the fully walled area to where the opening is above. It is these weak spots in construction where cracks are most likely to occur. The different weights of a full height wall and a low sill next to each other can cause problems if the footings are not strong enough. Bridging stones in the wall should also be placed below each side of the opening to avoid cracking in this location (Figure 6.5).

Figure 6.4: Cracking at a wall where loading conditions change, Bairnsdale, Victoria. Although large stones have been used, the crack may result from differential settlement.

Figure 6.5: A simple stone cottage near Bombala, New South Wales, with flat arches over small windows.

Windowsills made of a solid stone should be bedded with mortar at the ends only, as greater settlement at the sides of the window might otherwise crack the sill in the centre. The rest of the sill can be grouted underneath with a weak mortar after the stone is laid.

Damp-proofing

To prevent dampness rising through the wall from the ground, a damp-proof course (DPC) or waterproof layer should be placed just above ground level. This can be a plastic or bituminous membrane, a metal strip, or overlapping courses of second-hand roofing slate. If movement occurs in the wall, a brittle DPC such as slate may crack and let water through, so a waterproof additive to the mortar can be used as well. The DPC ideally should be at a height of 150 mm above the ground level. If ground level is above floor level a vertical DPC is also needed to isolate lower walls from the water table.

Weatherproofing window openings

Metal or other flexible flashings should be fixed into the wall return around window frames for waterproofing (especially for cavity walls), depending on exposure. At the sides of the frames, some windows have a groove where a flashing strip can be inserted. It is important to have some sort of baffle so the wind cannot blow through.

Figure 6.6: Owner-built house near Bairnsdale, Victoria, using flat stones, some laid on edge.

Windows are set in from the outer leaf of the stonework at each side of the window frame. This is more difficult to do in stonework than brickwork as it involves laying matching sized stones around the windows. If neither flashings nor setbacks are used at the sides of windows, silicon caulking is then necessary for weather-stripping to protect the building from rain damage.

Flashings for sills are placed directly under the window frame. If a solid stone sill is used, it should be shaped to catch drips from the timber windowsill sitting on it, and to carry the water away from the window. In this case a flashing under the stone sill can usually be omitted. Old stone sills often have a curved rim that sits under the walls on each side, and an inside edge rim up to the underside of the window. These are level, to build on, while the main portion of the sill is sloping to the outside. Sills made from a number of fitted flat stones should have waterproof mortar between them.

With these essential principles of loading and waterproofing established, we can now move on to the practical side of building stone walls (Figure 6.6).

Setting out

To mark the positions where the walls are to be built, set the site out with timber hurdles and string lines. Mark the hurdles with saw cuts to give dimensions for the

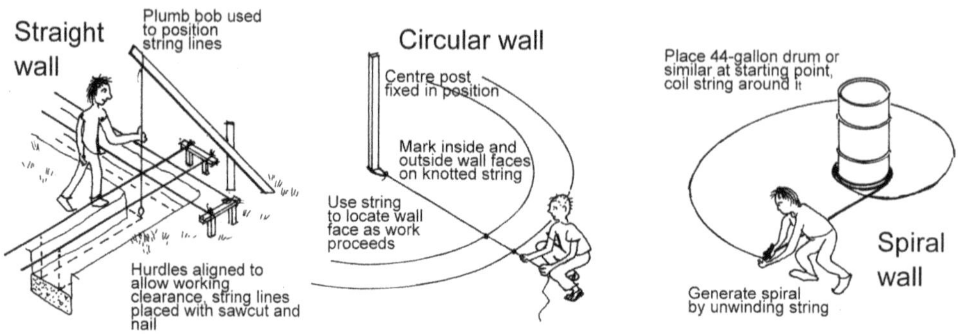

Figure 6.7: Setting out.

footing width and wall thickness, and then tie the string line to a nail on the required notch. Where the lines cross to mark a corner, a simple timber frame can be constructed overhead from which to drop a plumb bob, and the corner stones can be laid to that line (Figure 6.7). A spirit level against a strip of wood should be adequate to establish plumb on the straight wall sections as the stones are put into place.

In some designs, curves may be desirable, either for the effect or simply to avoid the need for corner stones. The curve is set out by placing a vertical stake at the centre point and marking the string line with a knot or a wire tie at the other end of the radius for the internal and external wall surfaces. A looped connection at the centre stake can be slipped upwards as the wall is built higher. A spiral curve

Figure 6.8: Pottery shards embedded in stonework where laid stone leaves a gap to be filled, Yandoit, Victoria.

may be generated by placing a cylindrical object, such as a 44-gallon drum, at the centre and taking the string line around it from a stake alongside.

A short stick cut to the same dimension as the width of the wall can be used to check the thickness of the stonework and keep it regular.

Laying the stone

The pattern of the stones laid on the wall depends on the type of stone available and how much shaping is done to each stone before it is laid. For house construction it is usual to lay the stones as found or quarried, with only a minimum of shaping to help fit the stones into place. Mortar is used to bond stones and provide a bed for each stone so that a perfect fit is not needed between the stones (Figure 6.8). It also fills the gaps so that air cannot flow through, and takes up space, so a wall using mortar needs fewer stones than a dry stone wall. Any type of stone can be used to build a rough stone wall of this kind, which is referred to as rubble stonework. The corner stones may be rough also and laid in the wall as found, or some shaping may be done if the stone can be worked. Corner stones or quoins are usually larger than the other stones. If the wall stones are too irregular, it may be preferable to use cut or worked stone of another type, or bricks, for corners.

The thickness of a rubble stone wall should be greater than a similar wall in brickwork, due to uneven fit, and because the rough rocky sides (perpends) reduce the bearing area of the stones on those below. The thickness of the joints depends on how well the stones are fitted together. Shaping the base, top and sides of a stone can reduce the thickness of mortar joints. Shaping the face of the stone is usually unnecessary except to avoid rough projections in internal walls.

Rubble walls

Rubble stonework is usual for house construction. As less time is spent shaping stones, these walls can be built more quickly than ashlar stone walls. The most shaping done on rubble stones is to trim around the face of the stone to give it a roughly rectangular shape while the sides built into the wall may be at all sorts of different angles.

The first concern when building with stones of irregular shape should be the strength of the wall (Figure 6.9). Ideally, the stones should have a flat base and top, but if this is not possible they should be laid with the weight leaning towards the centre of the wall, not outwards. As stones come in different shapes and sizes it is seldom possible to lay them in as regular a pattern as brickwork, where each one spans the vertical joint between those below. This is the strongest wall pattern, as the weight of the masonry distributes evenly downward to the base of the wall.

Figure 6.9: Granite addition with round opening, near Benalla, Victoria.

While laying stone it is best to build the tops of adjacent stones level wherever possible, so a stone in the next course can span across them. Where the fit between stones makes it difficult to avoid a vertical joint three or four courses high, the stone above this joint should be large and placed centrally across it so that a weak spot is not created.

Coursed rubble: A very regular effect can be achieved by sorting the available stones according to the height of their face. They are then laid in horizontal beds so that each course has a consistent height but differs from those above and below (Figure 6.10). If there is too much variation between sizes and shapes of the available stones for them to be neatly sorted, and especially if there are more flat stones, a pair of stones of equal width can be stacked occasionally to make them up to the height of the other stones in a course.

Random rubble: Stones that do not present a regular squared face surface can be fitted together without continuous courses. They may be of all sizes and shapes, but occasional through stones, the full width of the wall, should be incorporated to add strength. If sedimentary stones are used externally, an attempt should be made to lay the bedding layers flat. These are more likely to suffer from weathering if laid vertically as high face stones. Layered metamorphic stones such as slate should be more resistant to weathering. These can be laid to good effect with the larger slabs of stone placed vertically among a wall of predominantly narrower stones laid flat.

Figure 6.10: Coursed rubble (top), random rubble (centre) and dry rubble walling (bottom).

If there is any doubt about the weather resistance of stones laid vertically, they should be used only in a sheltered position.

Dry rubble: Essentially this is random rubble laid without mortar. It requires more skill in fitting stones together than a cemented wall, but if done well it can be very strong. In old buildings, dry rubble walling was more typically used for barns than for houses. In some places the barn stonework has outlasted the accompanying mortared stone building once the roofs are gone. Today, dry rubble is more likely to be used for garden walls (see Rainsford-Hannay 1957 for a description of the traditional art of dry stone walling). The top course or coping may have its stones laid vertically (like books on a shelf), so they are less likely to be knocked off, or they may be bedded in cement, to prevent rain damage as well as to bind them more securely. Clay or shale fragments may be used to fill gaps between stones, but should not be used structurally or cause stones to be spaced further apart, as clay or shale can wash away. Some through stones should be placed at regular intervals, halfway up the wall height, and near the top just below the coping stones.

Figure 6.11: Random rubble built to courses.

Random rubble built to courses: Irregular stones are used, but they are fitted so they come up to a level bed at intervals two or three stones high (Figure 6.11). The position of these intervals is usually determined by larger stones at the corners. The corner stones may be squared in contrast with the rough wall stones. Flint walling uses much smaller stones (Figure 6.16).

Squared or snecked rubble: Although stones of many different sizes are used, their faces are squared to give the wall vertical joints and level beds. Smaller filling stones (also squared) are used to equalise the level between adjoining stones, but courses are not continuous (Figure 6.12).

Squared rubble in courses: Stones of irregular size are used, with their faces squared. They are brought up to horizontal beds periodically, in the same manner as coursed random rubble to level up with larger corner stones.

Polygonal or rag walling: Stones of any shape are set in the wall to fit among other irregular stones. The beds are not necessarily horizontal and the stones are not laid in straight courses (Figures 6.13 and 6.15).

Bookleaf walling: Stones of about 75 mm or less are laid in regular courses. The length of stones may vary considerably. Stones of contrasting colours are often chosen.

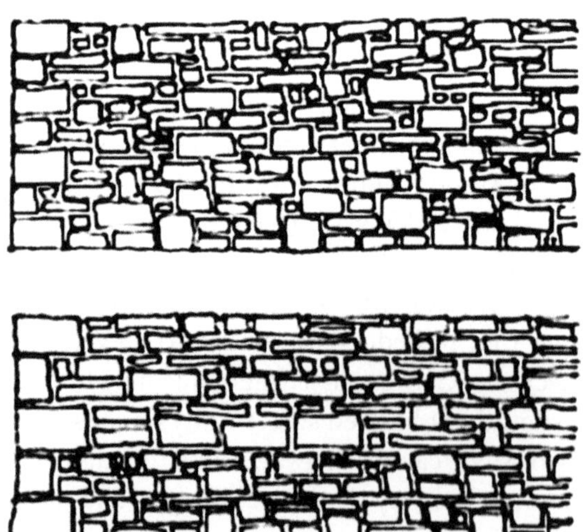

Figure 6.12: Snecked rubble (top), squared rubble in courses (bottom).

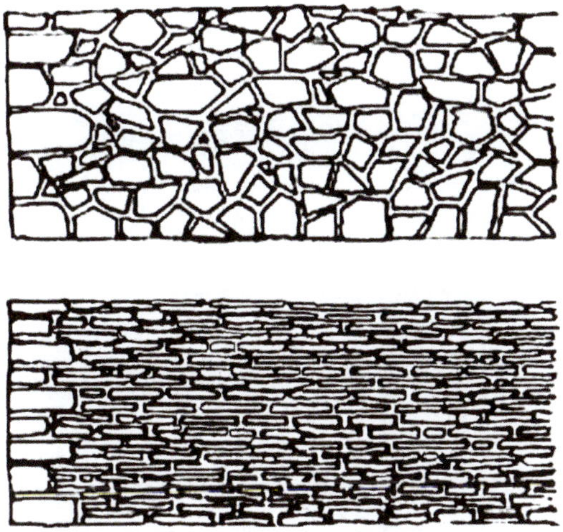

Figure 6.13: Polygonal rubble (top) and bookleaf walling (bottom).

Measuring

A folding metre rule that fits in a pocket is ideal for checking dimensions between the wall section that has been built and the next stone to go on it. Otherwise, when finding the next stone it often feels comfortable to use hand measurements such as

Figure 6.14: Carved record of construction date, farmhouse near Bairnsdale, Victoria.

Figure 6.15: Polygonal walling in a spacious porch, country house near Bairnsdale, Victoria.

Figure 6.16: Flint walling with brick corners and render for sill and buttress capping, Carapook church, Victoria.

a span, a fist or how far up a finger the next stone needs to be, rather than finding a tape or ruler all the time to check for best fit. The length and width of the trowel can also serve as a handy measuring tool at times.

There is always a bit of leeway available when building rubble stonework because of the variety of shapes and surfaces of the different stones, but it is better to find the right sized stone than to trim off one that is not quite right, at the risk of breaking off too much. There are probably only a few stones handy that are just the right size and shape for any one position in the wall, to span across the joint below or to fill in between the other stones. If there aren't enough of the right stones to choose from, it may be time for another trip to the quarry (unless the wall is nearly finished and what is left can be shaped to size). The best and most careful work should be in the most visible location, the walling at the base and the top will not be seen as much.

Ashlar

Unless the available stone is very easy to shape, ashlar is labour intensive and costly to use, so not really practical in most cases. However, ashlar dressing stones may be used as features in conjunction with rubble stone (or brick) walls. Wrought stone is ideal for corners, sills, arches or lintels and copings (Figure 6.17), to give the quality a bit of a lift. However, to build a whole house in ashlar would be prohibitively expensive. Internally, facing stones or veneers may be used for feature locations.

Generally used for public buildings, ashlar is built with sawn or accurately worked squared stones. Each stone has five worked surfaces: top, bottom, sides and face, or six if it is the full thickness of the wall. Joints in ashlar may include stone or cement joggles, dowels, or cramps (metal or hard stone).

In ancient times, bronze cramps were used in the Middle East and silver or gold cramps were used in the Americas. Iron is unsatisfactory as it corrodes unless well bedded in concrete. Sometimes the stone is connected with a mortise and tenon shaped in the stone. The joint lines on the face of the wall are then generally filled with a non-structural grout rather than conventional mortar.

More commonly now, stone is dressed into blocks with a rectangular wall face and set in cement mortar. The thickness of the mortar depends on how smoothly the sides of the stone are worked. Ashlar is often only a facing, with a backing of rubble stonework, concrete or brickwork (if used with brick, stones should be sized to match brick dimensions, especially height, with allowance made for brickwork expansion).

In ashlar walls, the stone face is worked and may be finished in a number of ways with the hammer and chisel, forming a pattern on the surface with the chisel marks, or it may be left with a rough rock face. To remove the chisel marks and create a smooth finish the face of the stone can be rubbed with stone and water.

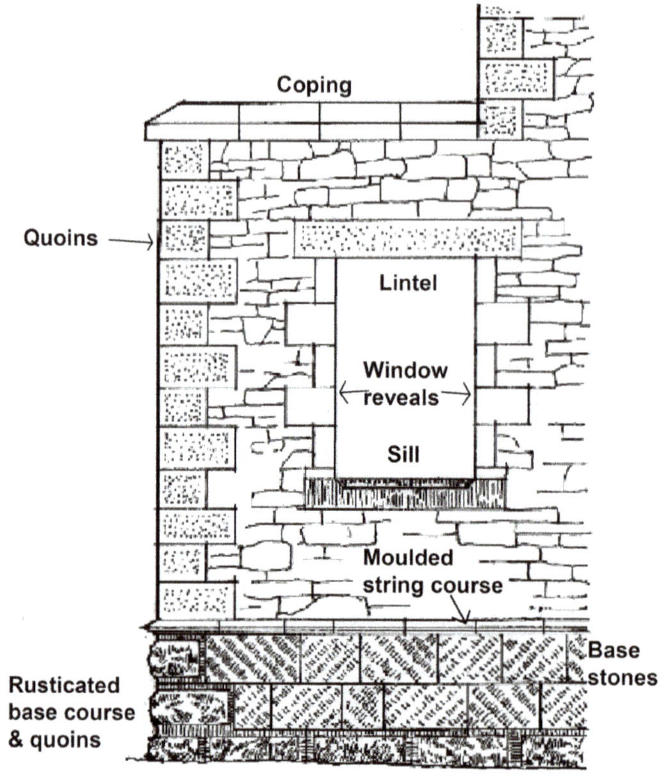

Figure 6.17: Dressing up a rubble stone wall with ashlar.

Mortar

Before mixing a batch of mortar it is best to prepare some of the next stones ready to go on the wall. There is an advantage in working progressively from the bigger stones, which need to settle on a thick bed of mortar, and the smaller or narrower stones that will fit between them. If this pattern is followed consistently, the diverse stones available will be used in the same proportion in the different parts of the wall.

Depending on how much time is spent shaping the stones, the amount of mortar mixed can be enough to lay the prepared stones, or more can be mixed to keep going for a short while. Once the first stones are laid, others can be found that will fit between them to use up the rest of the mortar.

If a stone will sit in the right place without mortar then it should be stable over a long period of time once it has mortar around it. If a stone has a tendency to slide or does not sufficiently span a vertical joint between the stones below, the mortar should not be relied upon to make up this deficiency, as cracking is more likely to occur in these circumstances.

Figure 6.18: Stone with mud mortar at Benham Estate, Tasmania.

The mortar should be a binding material with sufficient strength not to crack or crush under concentrated loads. However, it should not be stronger than the stones in the wall, as cracking might then occur through a stone rather than around it, which is more difficult to repair.

The proportions of cement (and sometimes lime) to aggregate, the size of the aggregate particles and the amount of water in the mix can all affect the mortar strength. A very fine sand needs more cement to make a coating around every particle than does a coarse aggregate, whereas a coarse pebbly sand may prevent stones from sitting tightly together.

Alternatively, for those who like to use truly local material, clay or mud can be used between stones (Figure 6.18). Doing this properly requires as much care as a dry rubble wall. If structural strength is a requirement, an experimental wall should be built and tested first.

With clay or mud used as mortar, it is important that each stone makes a neat fit with the others and will not wobble, as the clay is more likely to settle when more weight is put onto the wall. Any movement will accelerate the weathering process. Cement-based mortar is stronger and more waterproof (though also brittle) and can be used to space the stones apart or to prop them up where the fit is not exact.

A greater proportion of coarse aggregate (gravel) in the mortar can increase its strength, and the larger particles support each other between the other stones. For

Figure 6.19: Old barn at Tuross Head, New South Wales, random rubble laid to courses, perhaps built with formwork.

formed or irregular rubble stonework some gravel may be added to the mortar to form a mix that is more like concrete for extra strength in the larger gaps between the stones. However, for normal horizontally laid structural stonework, the use of larger aggregate just causes the stones to be set in place with wider joints and uses a greater proportion of mortar (Figure 6.19).

Aggregates should be free of impurities such as clay, organic matter, salt or any other material that might reduce the mortar strength or react with the stone to accelerate weathering. The aggregate should be stronger and more durable than the rest of the mortar mix if it is to improve the quality of the mortar. Clean river sand with particles of varying size up to about 3 mm is good for laying irregular stones – this may be referred to as concrete sand. It is too coarse for brickwork.

The cement supply should be kept in a weatherproof location, as even the bags can harden with exposure to moisture in the air, if left for a long time between use. Sand can be left out in the rain, it just requires less water when it is used wet. However, sand left too long can collect windblown leaves and sticks, as well as children's toys, all of which need to be carefully screened out when it is used.

The proportion of water should make the mortar into the consistency of thick mud, without being runny, so it can be easily trowelled into the spaces between stones. Lime may be added to improve the texture and workability of the mix.

It is a good practice to wet each stone before it is laid on the wall, as this prevents it from absorbing moisture from the mortar mix. The mortar should be stiff as well as workable, because a wetter mix takes longer to set and is not as strong.

A cement mixer is useful for mixing a quantity of mortar quickly, however, a wheelbarrow is adequate. If the shovel is too heavy and awkward, a hoe can be handy for mixing mortar in the wheelbarrow.

Wrought stonework does not depend on the bedding material for strength as the stones are carefully fitted. The mortar generally used for bedding is mason's putty: 3 parts stone dust to 1 part lime putty (quicklime and water). The joints are grouted with either neat Portland cement or 3:1 stone dust and cement. White Portland cement may be used to avoid staining. White Portland cement is used in the bedding material for white marble and for marble wall linings, while plaster of Paris or Keene's cement is used for grouting.

Limestone is laid with a mortar consisting of 1 part Portland cement, 3 parts lime, to 10 to 12 parts stone dust.

Alternatively, a lime:sand mortar of 2:5 may be used (but not where high early strength is required).

Granite can be laid with a stronger mortar of 1:3 ratio of cement to washed sand or granite dust.

Sandstone may be laid with a 1:1:6 mortar ratio of cement:lime:sand, but for some sandstones lime in the mortar is likely to cause stone decay. If the same sort of stone is soft or shows signs of weathering in other buildings, a ratio of 1:4 Portland cement to finely crushed stone of the same type or clean sand is more suitable. The mortar should not be stronger than the stone. In some cases a weaker mortar of 1:5 cement to sand may be considered. Mortar for pointing may be slightly stronger but not stronger than the stone when set.

Colouring pigments may be added to the mortar. A variety of mineral oxide colours are available. They should not be used in quantities greater than 10% of the cement as this weakens the mortar.

Water used in the mortar should be clean and drinkable and free of impurities such as salt, acids, alkalis, oils or sulphur. Take care not to add too much water to the mortar mix, which should be stiff and not sloppy.

A thicker bed of mortar is needed under larger stones to support the extra weight. When the base of a rock has irregularities, a thick bed can allow for these, and lets the rock be tilted. The mortar bed is trowelled or shovelled on, and trowel marks made to groove the top so the mortar has room to move when the stone is pushed down. The face edge of the mortar should be smoothed off so that it does not spill.

Before each batch of mortar is prepared, the larger stones should be placed on the wall to work out the best fit. They are the first to be laid once the mortar is mixed, and the smaller stones and long thin stones are laid next, to level off or to

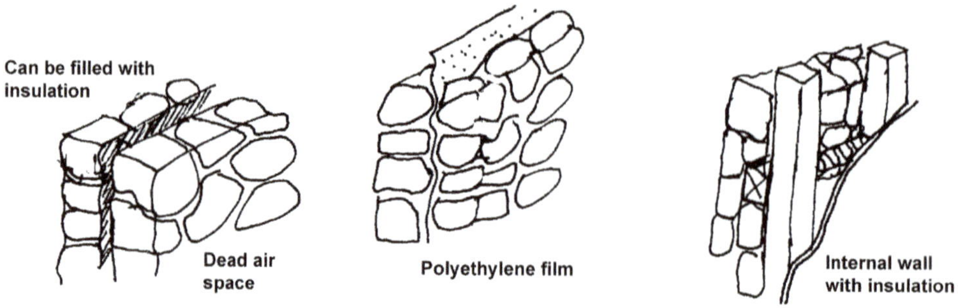

Figure 6.20: Insulating walls and moisture barriers.

span over vertical joints. As soon as the mortar batch is finished, the face of the stones should be cleaned with water and a brush or sponge. If left too long, the mortar drips can harden and become difficult to remove.

External walls

In wet climates, convention (or the local authority) may require a verandah around external stone walls, except those built with a cavity, waterproofed mortar or other barrier (Figure 6.20). Solid stone walls greater than 400 mm should be adequate against moisture penetration except in very exposed positions (Figure 6.20). External solid rubble stonework less than 400 mm should have 1 metre of roof overhang to ensure internal dryness.

In exposed positions where a wall is subject to wind-blown rain, cavity stonework or timber-framed stone veneer walls would be preferable. This allows any water that may seep through the stone to drip down the cavity and through weep holes provided between stones at the base of the wall.

Cavity stonework

The advantages of building a stone wall with a cavity are that moisture will not penetrate to the inside wall face through the mortar or the stones in the wall, and that the air space helps to insulate, slowing down the convective heat transfer through the stones. The inner and outer leaves of the wall are held together with brickwork ties or bent wire. Cavity stonework may also be built with brickwork or blockwork as the internal leaf of the wall. Weep holes (open perpends) are necessary between stones at the base of the outer leaf of the wall, which is solid below this level (at least one course from ground level).

Metal wall ties should be spaced as for brickwork. Unlike brickwork, the surfaces in the cavity of a stone wall are irregular. Approximations may be needed, and some stones may extend almost across the cavity. If some stones are as wide as the wall, and not porous, they may be used as through stones across the cavity instead of wall ties. These should be placed about 900 mm apart every three or four courses.

Figure 6.21: As explained by the owner builder at South West Rocks, New South Wales, this stone veneer building method was chosen because it is cheaper 'than paying a brickie to build it in bricks'. The stone was sourced from a local council quarry.

Stone veneer

A stone veneer to a timber stud wall can be built with less stone than a free-standing stone wall (Figure 6.21). A stonework thickness of 150 mm (6 in.) should be adequate for rubble stone, less for cut stone. The insulating qualities of the wall can be improved with reflective insulation fixed to the studs in the middle of the wall. Insulation batts can also be placed between studs, although the heat storage capacity of the stone is not as fully utilised when insulated from the inside of the building.

The stone veneer should be tied by metal wall ties to the stud walls at intervals of no more than 460 mm × 610 mm, based on brick dimensions (an average of 530 mm each way).

Formed masonry

Formwork can make it easier to build straight, plumb walls with rough stones. Stones can be laid to a height of about 600 mm per day, and the formwork is moved upwards when the stone mortar is firm.

There are various kinds of formwork that can be used. Timber planks can be placed on a fixed framework in the mosaic rubble construction method (Flagg 1922). Where a large number of walls are to be built, more permanent sets of formwork can be made up. The number of forms needed depends on the size of the building.

Two forms are used one above the other on each side of the wall and spaced apart with wires. Theoretically, eight forms are necessary, four for each side at the corners, but more are needed if a larger amount of work is to be built at one time. The stone is placed in the formwork and mortar is poured in between. As

Figure 6.22: Formed stonework at Tubbut, Victoria, using Flagg's mosaic rubble method. Even with the irregular stones, the corner has a straight edge. Windows have a rendered reveal. It is important to remove forms promptly in order to clean face stonework.

soon as the forms are removed, excess mortar should be scraped from the stones and the stone surfaces cleaned down. Forms must be removed before the mortar is too hard.

A simpler set of formwork is used if the walls are formed on one side only. With this technique, the stones are laid conventionally for one face, and set up against the form on the other.

Formed stonework is preferred by some people over traditionally laid stone walling, for its economy of construction where there is a repetition of similar elements (Figure 6.22). To be most efficient, the formwork should be in modules suited to the design, each of which is a few courses in height. It should be totally demountable. The number of forms is minimised if walls are kept to a minimum number of modular lengths.

Unlike laid stone, which can be built a few stones at a time, the stones must be available and ready to go on the wall for the whole height of the formwork. The stonework has to be built quickly for each lift of the formwork, so the mortar will still be fairly fresh at the bottom. The forms must be removed within a day or two so that the mortar can be cleaned off the face stones with a wire brush, scrub brush and sponges. If left much longer, the hardened mortar is very difficult to remove.

For its strength, formed stonework does not rely so much on the placement of stones (which for laid stonework are best in an orderly flat configuration) as on the concrete poured into the wall behind and between the stones. Reinforcing is generally placed in the centre, especially if the shape of the stones makes them incapable of transferring much structural loading.

Flagg's mosaic rubble

The *mosaic rubble* method was outlined by architect Ernest Flagg in the 1920s (see Flagg 1922). The procedure involves a number of repetitive steps, to be done by a team of workers provided with the correct formwork. Flagg claimed his method was straightforward with these advantages:

- the work can be done by unskilled labour, and
- the formwork construction involves no nailing and no waste.

Footing: The footings are finished off level and damp-proofing placed (at the underside of a concrete floor or 50 mm below a timber floor). Module lines are marked on the surface of the footings.[4]

Sleepers: Tapered and greased timber cross-pieces called sleepers are laid across the footings at half-module spacings. The sleepers have holes drilled each end for inserting steel pins at the base of each upright. During construction, the sleepers remain embedded in the masonry wall but are easily driven out when the stonework is completed.

Uprights: 100 mm × 100 mm timber uprights are pinned to the sleepers. Corners go up first, temporarily braced to stakes driven into the ground. The uprights have holes drilled through at 150 mm intervals to fit toggle pins (short, sharp pointed spikes), which will keep the formwork at the required height once stonework proceeds. When corner uprights are in place, the other uprights are pinned on the sleepers. The tops of uprights are held in place with alignment trusses to match the wall width, held by wires and pegs, braced horizontally. Diagonal bracing is fixed to corner pairs of uprights. Alignment trusses are rectangular frames fabricated from 100 mm × 50 mm beams with cross-bracing in the end bays.

Formwork: Planks 150 mm high are placed two or three deep, first on the sleepers, then on the toggle pins as work proceeds. The form planks are placed with 25 mm × 25 mm release sticks against the uprights, held apart with spreaders. At about every third plank, where needed, wires are looped around the uprights and twisted to hold the spreaders tight against the planks. The stones are placed inside the formwork, the flattest side flush with the outer face, and concrete is shovelled in between. The wires can be left in the wall and their projecting ends cut off.

Re-using planks: When stonework is built to the top of the upper form planks, the lower forms – where the cement has set already – can be moved upwards. Before the release pins are knocked out, toggle pins are tapped into the holes in the uprights to support the upper planks. The lower planks are taken out daily as the work proceeds, and the uncovered face stonework can be cleaned and pointed. The technique recommended by Flagg is to set fine stone chips into the gaps, instead of smoothing or pointing the joints in the ordinary way.

Upper walls: Interior uprights are removed when upper floor beams are set in place. They can be re-erected, to continue the stonework at the upper level, resting on the floor beams. On the outside, the posts are left in position, held by wires in the wall. Upper posts are then positioned directly above the lower posts, fixed with a vertical pin, similar to the pin holding the lower post to the sleepers.

Openings: Window and door formwork must occupy the full width of the wall between the planks. The forms should be splayed so they can be easily removed once the planks are taken out. The door and window frames are set in place with the forms. Each frame has permanent wall ties to anchor it and hold it in position.

Scaffolding: Supports for scaffolding planks can be bolted to the uprights holding the formwork.

The finished wall surface of formed stonework is relatively flat and true to line. Interiors may be bagged or rendered if the natural stone finish is not preferred. Alternatively, plasterboard may be placed up to the formwork of the inner wall face. The concrete infill in the stonework fixes it to the wall.

Framed formwork

In their classic text on self-sufficient living, Helen and Scott Nearing described their own stone construction process, which they based on Flagg's mosaic rubble technique (Nearing and Nearing 1970). Their method differed in that they made up forms 1825 mm (6 ft) long × 410 mm (16¼ in.) high, more bulky than Flagg's planks. These were built from white pine or spruce because of its light weight. The Nearings used a concrete mix of 10 parts gravel to 5 parts sand and 1 part cement.

Slipforms

The *slipform* method of formed stonework is a slightly different technique, developed by Karl and Sue Schwenke for use in their stone house, and based on the Nearings' method (Schwenke and Schwenke 1975).

The slipform is made up to a size of 8 ft (2.4 m) × 19 in. (475 mm), being the maximum size that Karl and Sue Schwenke could comfortably handle. It is framed with 100 mm × 50 mm top and bottom and form studs at 600 mm centres. Facing boards are applied, which are 25 mm thick and have a 12 mm gap between them. The gaps are used for wiring forms together across the width of the wall. A timber spacer is used to keep the forms the right distance apart.

Because the slipforms are framed up with short studs, they have reasonable stability and can be bolted together. Bracing is fixed to the forms, and the bored posts used in Flagg's technique are unnecessary. However, removal of forms once each pour has set is not as simple as the removal of Flagg's plank forms, especially at the base of the wall where leverage cannot be used from below.

One-sided formwork

The use of formwork only on the inside face has the end result of a regular, flat inside surface, while the outside of the walls remains irregular with perhaps the occasional protruding stone, a more traditional stone appearance (Figure 6.23). An Australian example of building this way is documented by Jim and Anita Maxlow about the construction of their own house (see Maxlow and Maxlow 1984).

Services

Plumbing pipes and electrical wiring can be built into stonework, and these need careful forethought, because making holes through stone walls is difficult work if positioning of services is not planned in advance.

Copper or plastic water pipes can be built into the wall with mortar, but their couplings should be tested beforehand, to avoid the possibility of discovering leaks after they are embedded in the wall. Small pipes or electrical conduits can be placed along a deepened course joint where it is built long and straight enough, or ducting space can be left on the inside of a wall. To take pipes from one side of a wall to the other, duct holes should be left for sinks and other plumbing units, and for water supply and electricity.

Figure 6.23: Local stone addition under construction with single-sided formwork, plank forms set in place on support posts, Daylesford, Victoria. The narrow vertical returns of some corner stones indicate that they rely on the concrete packed in behind as much as the bedding.

Since it is sometimes difficult to predict future needs, it is a good contingency measure to leave a gap somewhere in the wall between all rooms with a wooden pole or block as a spacer. If it is not needed, the gap can be filled later with a weak mortar.

Electrical outlet boxes can be built into the wall, fitted between stones, and the conduit can be placed between them and wires laced in afterwards, or wires can be laid in place as the required height of the wall is reached. Similarly, allowance should be made for telephone connections in rooms where they are required.

Finishing joints

The bedding joints and perpends of the stonework should be given a consistent finish. A flush joint in line with the stone face is appropriate for internal wall faces where a smooth finish is preferable. If the stones are irregular this may be difficult to achieve.

A struck joint is made by running the trowel along the joint before the mortar is hard, either pressing in at the top: upstruck (or weathered), which exposes the mortar to the weather, or pressing in at the bottom: downstruck (or struck). The downstruck joint exposes the top edge of the stone to the weather and is preferable for Australian conditions.

Raked joints give a good shadow line and help the stones to stand out from one another. For an irregular wall face this works well. The mortar is raked out to a depth of 6 mm to 12 mm behind the stone face. The stone should be of sufficient hardness to avoid weathering because of the hollows formed, where water may accumulate. Round raked joints or keyed joints are made with a round jointing tool before the mortar sets. This forces mortar from the centre to the edges of the joint, and can fill any irregular edges around the stones with mortar.

Tuck pointing is the application of an applied marked joint, usually to make the work look more regular than it really is. First the joints are raked out to a depth of 25 mm, then they are cleaned out and, in the case of old work, well wetted to reduce drying too quickly from suction. They are then filled with a pigmented mortar to match the stone colour, flush with the stonework (Figures 6.24 and 6.25). Finally, when the mortar is semi-set, the joints are pointed with white lime putty in straight lines, which is trimmed and the stonework around it cleaned immediately after tooling.

Cleaning

At the end of a day's work the stonework should always be cleaned down. A scrub brush and a sponge should be adequate to clean off mortar with water, from the top of the wall downwards. Stubborn drops of mortar can be tackled with a wire brush.

Figure 6.24: Historic stone house, South Australia.

Stonework that has not been cleaned down as the work progressed can be washed with a solution of hydrochloric acid (10% in water). The cleaning solution should be washed or hosed off within an hour of use. Hydrochloric acid may change the colour of the stone as it brings out iron oxide, giving it a rust-like appearance. It is therefore important to wash with water soon after finishing each section of stonework. If left too long, mortar stains are difficult to remove except with chemicals.

Figure 6.25: Detail of historic stonework with tuck pointing including an unrelated line of white mortar marking a perpend where there is none, across a stone face.

Figure 7.1: Back door and windows, granite walling edged in sandstone, Abercrombie House, Bathurst, New South Wales.

'In stonemasonry, the weight of the stone is used to hold it up in the air, by the same force that would otherwise make it fall.'

Attributed to Charles Perrault

7

Upper wall construction

Building upper levels in stonework brings new challenges. A stone building does not easily compare in scale with a natural cliff formation. Yet, with art and artifice, a high stone wall regains a kind of natural power while still clearly the work of man. Quality stone and the ability of the stonemason can raise a wall to an imposing height.

Up to the ceiling height of the ground floor, the structure of a stone wall is fairly straightforward. The main problems are finding or shaping corner stones or making sure the rocks fit well together and are stable. In single-storey buildings, windows do not need to have stonework built above them, as they can be placed where they almost reach the ceiling, with the roof supported by timber lintels.

A two-storey stone wall requires more forethought. The lower-storey wall needs to be thicker to support the extra weight, and doors or windows need an arch or lintel overhead that is strong enough to support the stonework directly above them. The placement of doors and windows must be worked out so that there is enough stonework between openings in the lower wall to support the upper wall (Figures 7.1 to 7.3). Ideally, the doors and windows should be vertically aligned to avoid a line of weakness in the joints between the openings. Cracking is most likely to occur as a result of badly positioned openings.

The wall distance between openings should be at least as wide as that which would be needed for a free-standing stone pier, whether the openings are at the same level or not, so that the wall has enough stability around each opening.

Wall thickness

The wall thickness for rubble stonework should be at least a third greater than a similar wall in brickwork. So for a single-storey load-bearing wall using cement

Figure 7.2: Small backyard castle using a variety of stones and rendered finishes, near Launceston, Tasmania.

Figure 7.3: Interior timber lining of stone arch.

mortar the minimum wall thickness should be 300 mm. For some stones it may be difficult to keep the wall as narrow as this. It depends on the size of the stones and how well they fit together in the middle of the wall when face stonework is required on both sides.

For a lower-storey wall supporting a load-bearing wall above it, the wall thickness should be greater again by about 100 mm. Having a thicker wall below each consecutive storey can provide a convenient ledge to support the upper floor joists. The ledge need not be perfectly level, but should be brought to level in the positions where the joists or bearers will be placed.

Building higher up creates difficulties such as how to lift the stone up, safe scaffolding and stable wall construction. There could be problems getting a building permit for stonework beyond two storeys until the building surveyor is satisfied with the strength of the laid stonework. However, a cellar or small amount of extra stonework, for an attic or a parapet, is not unreasonable. The wall thickness needs to be planned from the top down, to allow for each storey change (Figures 7.4 and 7.5).

Brickwork	Stonework	Storey
(110 mm non-load-bearing) 230 mm	(150 mm non-load-bearing) 300 mm	1 of 1 (ground) 2 of 2 3 of 3
230 mm (to 7.5 m max. height) (350 mm beyond 9 m length)	400 mm	subfloor of 1 1 of 2 (ground) 2 of 3
350 mm (470 mm beyond 9 m length)	480 - 500 mm	subfloor of 2 1 of 3 (ground)
470 mm	630 mm	subfloor of 3

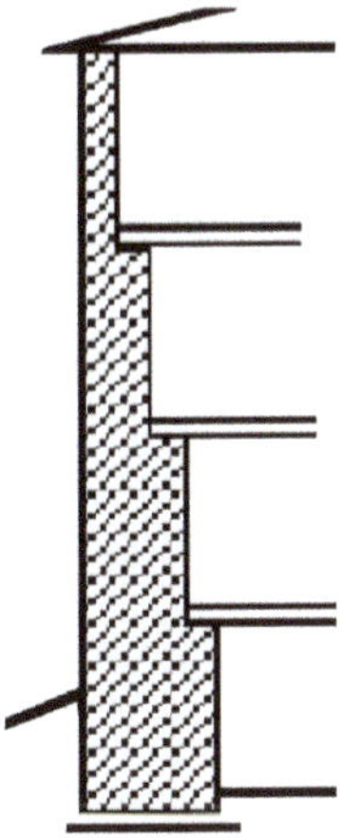

Minimum wall dimensions for masonry (cement mortar)

Figure 7.4: Wall thickness for rubble stonework.

Scaffolding

Building a two-storey stone wall generally requires scaffolding. For small patches of stonework it may be possible to build both inside and outside wall faces from an upper floor that is already built, cleaning off mortar from the outside with a ladder, but scaffolding is preferable. The combination of 44-gallon drums and planks is versatile for building one storey, but above that, post and beam scaffolding is better

Figure 7.5: Historic four-storey bluestone barn, Kyneton, Victoria.

Figure 7.6: Scaffold framing built like trestles, house near Hepburn Springs, Victoria.

(Figures 7.6 and 7.7). A scaffolding permit is required for two-storey work or a working platform higher than 2 metres.

As stonework is time-consuming, it is better to make up some scaffolding or buy it second-hand than to hire it. Bearing in mind that the termites will be

Figure 7.7: Plank scaffolding supported on timber post with bracing, suitable for one person working within a few metres above ground, near Hepburn Springs.

interested, timber can be used to reach modest heights. Old lengths of timber can be nailed or bolted together as a framework to support scaffold planks.

A scaffold can be partially supported on stonework that has been built already. Holes can be left in the stone wall for inserting scaffold beams that are framed into posts on the ground. (This is not recommended practice but sometimes it is done for small lengths of wall. Specifications can stipulate that this method not be used. Any holes left must be filled in later.) If the stone wall has progressed far enough above, and the scaffold is not too wide, beams cantilevered from a hole in the wall are sometimes used to support the end of a plank. Again, this can be risky and requires a very good understanding of the structure. The scaffold must be stable and strong enough to safely carry the weight of a number of large stones and a couple of people using it.

Lintels

The materials used for lintels are stone, concrete, steel or timber (Figure 7.17). Timber should only be used to support stone when it is a durable hardwood the thickness of the wall (or built up to wall thickness), and with the agreement of the building surveyor. Many old stone buildings still have their squared log lintels intact, and even fire damage should not cause such large timbers to fail, as the outside of the wood may char but the inside retains its strength. It is important that the stonework above the lintel is built to transfer the load to the

Figure 7.8: Rough lintel over doorway in hard local stone, near Cooma, New South Wales.

Figure 7.9: Three-storey Hobart stonework with dormers, double window with narrow stone mullion and three single windows neatly aligned on the right, with an extra small window centred.

sides of the opening in as few courses as possible, so that only the triangle of stone immediately above the lintel is directly loading it (Figure 7.1).

Stone lintels are the best from an aesthetic point of view (Figures 7.8 and 7.9), but there may be a problem finding a stone of suitable size for a lintel, and it is heavy work getting a large stone into place. In addition to the opening width, the stone should have a bearing on the wall of at least 110 mm at each end (Figure 7.10). Wrought stone lintels of granite or quality sandstone are sometimes used in conjunction with rubble stonework, if suitable stones for lintels cannot be found among the wall stones.

Concrete lintels may be chosen as a less difficult solution. They do not require the amount of lifting work necessary with stone lintels, as they are poured in place on the wall. They require formwork and reinforcing, and are most practical if the window sizes are standardised, as the formwork can be re-used. Either a concrete surface can be left on the face of the wall, or a thin veneer of stones can be placed against the formwork before the concrete is poured. The stones should not be placed at random but laid in a pattern that does not draw attention to it. For example, giving the veneer the look of a flat arch, with a wedged centre stone leaning on side stones, gives the impression of structural adequacy.

Steel lintels, such as those used in brick or block construction, are not as suitable for rough stonework, as the stones sitting on them must have equal depth across the width of the opening. This can cause problems if the stones are being used as found, and are too brittle to be shaped to size. If it is too difficult to find stones to fit over steel lintels, the use of bricks at lintels in conjunction with stone walls is a time-saving solution. Alternatively, steel plate may be used in the form of an arch bar (see 'Flat arches' below).

There are a number of different ways to reduce the load on a lintel. A relieving arch or a bull's eye directly over the lintel can take some of the load if a lintel is insufficiently strong, or in case a wooden lintel is burnt. The relieving arch may be immediately above the lintel (Figure 7.11) or a bit higher up. The stonework between the lintel and the relieving arch, or tympanum, can support the arch as it is built, thus avoiding the need for independent formwork, and allowing the arch to be built without smooth undersides to the stones that would be necessary in a clear spanning arch. The bull's eye requires more carefully shaped stones. It may be done in brick instead.

If a lintel is not quite long enough to have an adequate bearing on the sides of an opening, stone brackets (or corbels) on either side can be used to transfer the load, or reduce the effective span. This creates extra joinery work to fit into the shape, but between straight-edged stones the window frame can be adapted to suit.

The brackets may be used close together to form a sort of pointed arch over a narrow window. This is a very early construction technique and is not a true arch, but it can make a very effective detail. A monolithic arch is really a lintel not an arch. Its span is limited by the size of the stone used. It is not a common feature as it creates more work both in stone shaping and window joinery, and relies on the use of stone that is easily worked.

Flat arches

Flat arches or joggled lintels require the stone to be carefully shaped. The joints between the stones of a flat arch radiate outwards from a keystone in the centre, lining up with a point below, at the centre of the opening, although the base of the arch appears level. If the bottom of the stones is truly level, an optical illusion is caused, making them seem marginally lower in the middle. A slight camber in the line of stones can correct this as well as being better for strength. A flat steel bar, or arch bar, may be used as extra support for the stones, especially if they are not

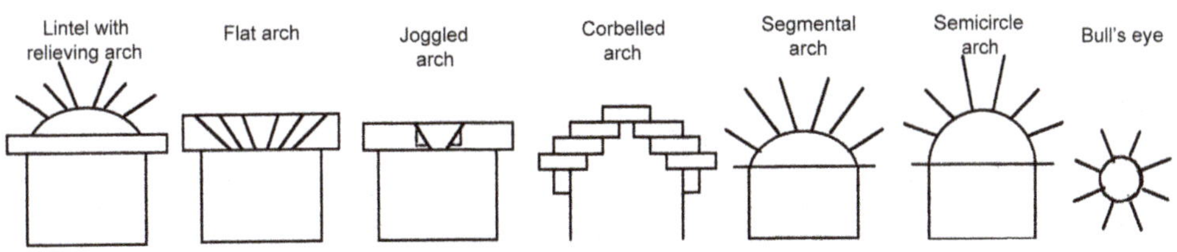

Figure 7.10: Lintel and arch construction methods.

Figure 7.11: Detail of door with relieving arch over stone lintel, Bairnsdale area, Victoria.

accurately worked and rely on the mortar for correct spacing. Bricks may be used in flat arches where they are cut to the correct angle. If used without cutting at angles for the camber arch, they are usually built in conjunction with an arch bar.

A joggled lintel is made up of three shaped stones. The centre stone is shaped as a keystone or may even be vertical at the sides, relying wholly on a secret key joint or joggle within the width of the wall. This joggle where the keystone meets the side stones prevents the keystone from slipping or the flanking stones from being pushed out of alignment (Figure 7.12).

Figure 7.12: The joins for the central keystone can be differentiated from the grooved lines representing joins in the two side stones of this flat arch, Cadman's cottage, Sydney.

Figure 7.13: Narrow window reduces number of stones in arch, Dubbo, New South Wales.

Arches

The curve of an arch is an ideal shape for openings in masonry, as all parts are in compression. However, the preparation work and fitting in of joinery for an arch is more time-consuming than for a rectangular opening (Figure 7.13).

Segmental, semicircular and other arches require formwork to set the stones in place during construction. A curved form or centre is placed to carry the load alongside the window joinery (which is not load-bearing), and the stones are laid in place from the springers up, until the keystone gives the arch its own stability.

To counteract the sideways thrust of the distribution of loading, the stone walls on either side must be wide enough to have a buttressing effect to hold the arch in place. The advantage of the semicircle arch over the segmental arch is that the curve brings the load to a vertical loading direction at the base. This does not mean the sideways thrust is absent, since the arch is not a rigid component like a lintel, but it means the forces of compression at the base are more resolved. The joints between the stones (or voussoirs) of a segmental arch may be shaped to radiate either from the centre that the arc is developed from, or from the height of the springing line at the centre of the opening. The rise of an arch adds to its strength in transferring loads to the sides of an opening. The downwards distribution of loading from one stone to the next is not as pronounced in a flat or camber arch, so more care must be taken with these to adequately support the stones at mid-span.

It makes the job easier if door and window sizes are standardised as much as possible, so that the same centres can be used for each size opening (Figures 7.27 and 7.28). If openings of varying widths are to be spanned with arches, they look best with their springing line at the same height. Unless there is a higher ceiling to accommodate a larger arch of the same proportion as a smaller arch, both should have the same rise. Typically, a semicircle arch is used for a narrower opening, matching the height of either a segmental or elliptical arch for the wider opening.

Joinery such as windows in the arch must have the same curve as the centre constructed to form the arch. The size and shape of the centre can be changed marginally by altering the thickness of the lagging strips around the curve, to adjust it for small variations in the size and shape of window or door joinery.

Above an arch, the distribution of loading should be balanced. The compression forces on the arch add to its stability. Upper level windows should align with lower level openings, to avoid uneven loading, as this can result in cracking. The position of upper floor loads and roof beams should also be considered.

Arch stones

Stones for an arch can sometimes be found roughly to the right shape among field stones or rubble stones, as long as the keystone has a good angle on either side. Rubble stone arches require properly shaped stones, either naturally or by chisel, making voussoirs that fit around the arch centre with enough of a gap between them for the required thickness of mortar.

All the stones for the arch should be fitted together before the arch is built, so that the angles can be worked out for the sides and for the inside surface. The stones must fit closely together, with adequate allowance for mortar, especially if the arch is to span 1 metre or more. Sedimentary stones should be shaped with the bedding layers roughly parallel to the joints between the stones, radiating outwards.

Arch voussoirs (like decorative stone lintels) can be made from wrought or machine-cut stone (Figure 7.14), even when the rest of the wall is built of rubble stonework. For continuity, if wrought stone is used for an arch in a rubble wall, the stones surrounding the opening, or at least for the sill, are generally of wrought stone as well (Figure 7.15).

Machine-finished stone can be bought from commercial stone merchants. For precision detailing, a drawing should first be done for each stone, which is numbered, showing its size, angle of cut and where it will be set. Once they are cut, some hand work may still be needed to ensure the stones fit exactly. The joints between wrought stones are small and evenly spaced. The bonding surface of each

Figure 7.14: Smooth coursed stonework, neatly turned into cranked voussoirs over arch, Dubbo.

stone is first wetted then dipped in a mortar with a fine sand aggregate, then it is set on the arch in its place.

Wrought stone voussoirs or other dressed stones are the craft of skilled tradespeople, unlike the rubble stonework. A stonemason experienced in working

Figure 7.15: Double hung arched windows in a curved wall, Narrapumelap, Wickliffe, Victoria. The stones for decorative returns and around the curve are of Waurn Ponds limestone, softer than the basalt walling.

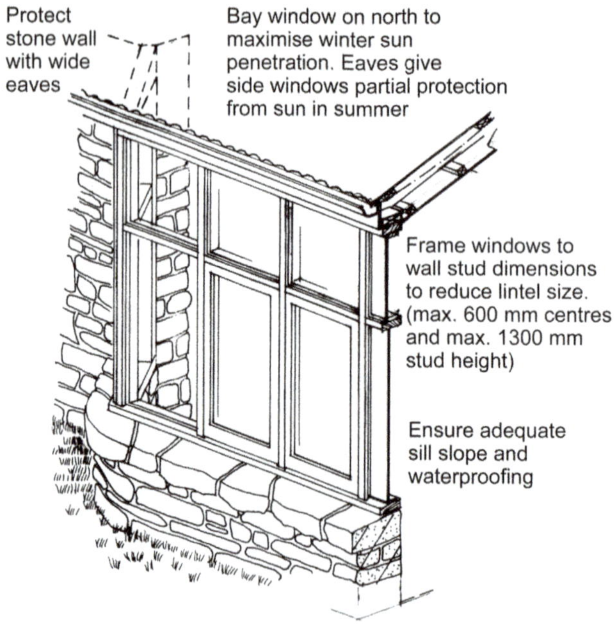

Figure 7.16: Sunny winter windows, sheltered stone walls.

with cut stone should be familiar with the jointing and grouting techniques of ashlar, or the methods that are usual for a high standard of finish, if this quality of work is required.

Windows

In two-storey work, or where a parapet extends above the roof, the exposure of the wall to weather creates opportunities for rainwater to get in.

Under eaves, the stonework is protected (Figure 7.16). In simpler construction it is possible to avoid the need for stone lintels or arches altogether. Windows can be placed directly under the roof structure if there is no need for stonework above, in which case the construction of the lintels can comply with timber framing techniques.

Dormer windows

One way to build in two storeys without the need for large areas of stonework on the upper level is to build the stone walls up to the next level just for dormer windows (Figures 7.17–7.19). But first a word of warning, the cost of an attic storey must be justified for other reasons. For the small area provided there is also the cost of an access stair (or ladder if it is only storage).

Figure 7.17: Stone-faced dormer window in line with the wall, Bombala, New South Wales.

The dormer window works best if the roof slope is fairly steep. The external stone wall on the sides of the window itself needs to be wider than a conventional timber-walled dormer, so that there is a stable stone column on each side of the window to support the lintel and the triangle of stones above it.

Figure 7.18: Attic interior with dormer window.

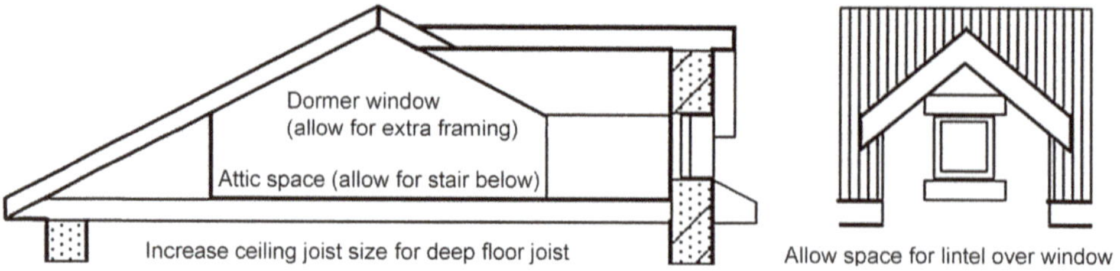

Figure 7.19: Stone wall to dormer window.

Wall flashings and drip moulds

Flashing is needed to direct the movement of moisture around the wall construction, to make it flow away from the building wherever possible. This includes all junction points where moisture could otherwise seep through the wall. Flashings are needed above roofs or upper level decking (Figures 7.20–7.22). Flashing around window openings has been outlined already. In addition, it is advisable to include flashing for any lintel of a denser material than the exposed stonework above it, such as brick, metal, concrete or a harder stone. Where appropriate, weep holes are left between the stones immediately over the flashing.

Figure 7.20: The language of these dormer windows emphasises the larger windows underneath while the dormer windows themselves are only small round openings each embellished by a decorative gable end and a curved drip mould, Abercrombie House, Bathurst, New South Wales.

Figure 7.21: Round turret with pointed roof and crown parapet, Abercrombie House.

Figure 7.22: Tudor arch to grand entry, with drip mould, flanked by narrow gothic arches, and gothic decorative windows above, Abercrombie House.

If a metal flashing is difficult to insert, such as above arches, a line of second-hand roofing slate can be built into the stonework above to direct water away from the opening.

In ashlar work, a moulded string course can be added to the top of a lintel or arch to deflect rainwater from the opening, especially if it is spanned with soft stone. A drip mould profile stops the flow of water by making it drip clear of the wall face.

Arched windows

Arched windows are difficult to find, and can be more expensive than rectangular windows if bought new, especially if the arched part is openable.

The least expensive is a fixed-pane arched window above an openable rectangular window. Some manufacturers make these as a standard item, and they are also the easiest for the owner builder to make (Figure 7.23). To simplify the woodwork for the glass, two pieces of wood can be cut for the inside and outside of the curve, one wider than the other to make a rebate on the outside of the curve.

Figure 7.23: Timber curve cutouts for an arched window with fixed glazing.

The two pieces are then joined. This makes it possible to use wood thin enough for a jigsaw to cut. The glass and putty fit into the rebate formed, and the top join is not seen under the stone arch.

The curve of the window joinery must be made to the same template as the arch centre. The centre is set up within the opening after the window is in place, and it should carry the load of the stones so that no stress is put on the window (Figures 7.25–7.28).

Second-hand arched windows (Figures 7.24 and 7.25) are not commonly available (understandably – after all, who would want to see any more beautiful old arched façades pulled down?). Many of the double hung window frames in older buildings have a different inner and outer shape, squared on the inside, arched on

Figure 7.24: The author's mother with a window found second-hand.

Figure 7.25: Windows for three levels of the tower, lintel at base and arches over, house near Hepburn Springs. Arched windows spanned with local hand-quarried stone.

Figure 7.26: Shaped sandstone, some from recycled sources, around recycled round window.

the outside.[5] This allows the bottom windowpane to be opened as high as the top. This kind of window can be placed close to ceiling level, where the upper storey wall is thinner than the wall at window level. An arch built to match the outside of the window can carry the upper wall, and a timber lintel over the inside part of the window only needs to support the upper floor. Double hung windows with a full arch frame at the top have the disadvantage that the bottom sash will not open all the way up.

As second-hand arched windows show up occasionally, it is worth looking around demolition yards and building antique dealerships as soon as the decision has been made to use arches, in case something special can be found.

This search should be done during the design stage of the building, as the end result of the design may depend on the nature of the find. Besides, if stone builders don't use the few arched windows that are around, they may end up in timber-framed buildings where the curved shape is unnecessary structurally.

Casement windows may be made with a full height openable arch but they must be flush with one face of the wall to open away from the curved soffit of the stone arch. Care must be taken with the water seal of any windows or doors opening inwards.

Arch centring

To build an arch, a centre is used to form it, since the window frame is not strong enough to support the arch stones. The centre is a frame with two curved ribs, made of timber, supporting lagging strips to hold the stones.

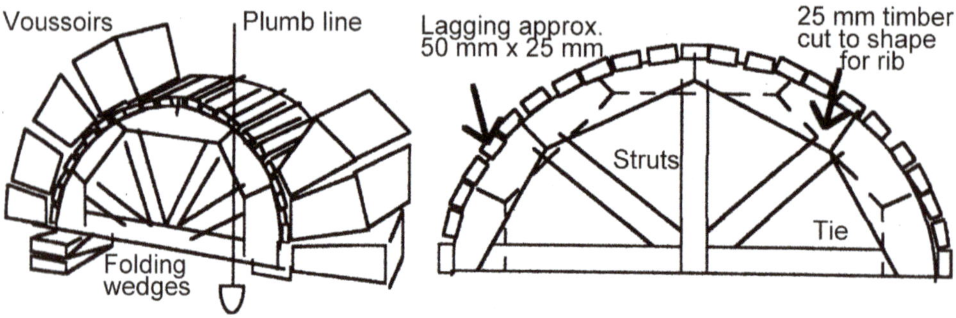

Figure 7.27: Arch centre.

Two ribs slightly smaller than the curve are cut, allowing for the thickness of the lagging strips, which are nailed onto the ribs once the centre is framed up. The top surface of the lagging should finish marginally wider and higher than the window frame, unless the frame is recessed behind the wall.

The centre is placed up against the window, within the thickness of the wall. It is brought to position with wedges, so it can be removed when the arch stones are

Figure 7.28: Arch centre in place outside window frame with arch stones laid, near Hepburn Springs.

Figure 7.29: Stone vault in retaining wall with decorative low garden wall, Abercrombie House, Bathurst, New South Wales.

all laid into place. The arch stones do not need to be in perfect radiating lines, as long as the stone at the top, the keystone, has a good wedge shape.

Vaults and domes

Structural stonework for a ceiling or roof is unusual in house construction, but stone vaults or domes are traditional in some parts of the world. If built today, they would be a special feature (Figure 7.29). Stone can be used instead of concrete for cellar construction, or for an earth-sheltered room. Any stone that can be shaped to the required profile would be suitable for the construction of vaults or domes. However, it is not an easy task. To be self-supporting, the curve must have adequate height.

This is the principle (Figure 7.30) which assumes the use of an easily worked stone. Achieving the required result would take a lot of dedication to the task.

The centring technique for a dome is to build up a series of radiating ribs. A thin strip of wood is fixed central to each rib generating the curve of the inner surface of the dome. A space is allowed at each side of these strips so lagging pieces of graded lengths can be fixed across the ribs.

Centring for a vault is much the same as for an arch, except that if two vaults meet, another rib is required on the diagonal (in plan).

The stones for vaults or domes should be accurately shaped (Figure 7.31). The required angles for each surface are taken from detailed drawings showing the plan

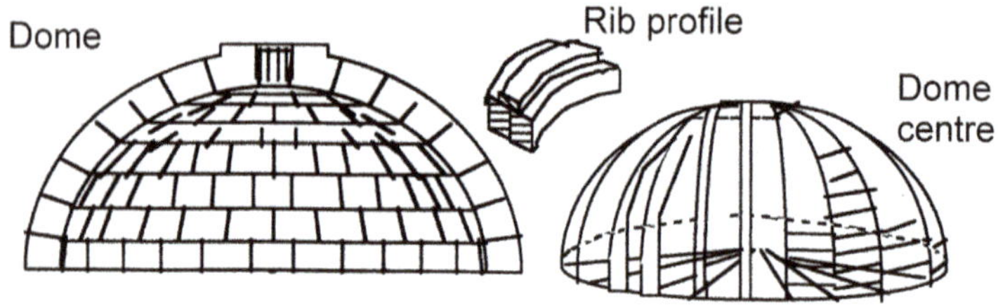

Figure 7.30: Dome construction.

and elevation of the stones for each layer of construction. This work should only be done with good stone. Precise structural stonework is a specialised skill. The development of computerised stone-cutting equipment has the potential to create further possibilities in this area.

Roof fixing

For rough rubble stonework without through stones, a concrete tie beam may help to bond the top of the wall and keep it from spreading under roof loads. However, this should not be necessary if the wall has been carefully laid to ensure stability across its width.

For nailing the rafters on, a timber top plate is laid along the levelled off top surface of the stonework. It should be placed within the middle third of the wall thickness for best stability. Steel pins or bolts are built into the stonework for the last few courses of the wall to a depth of 300 mm or more to hold the top plate firm and prevent uplift of the roof in strong winds.

If the roof slope is steep, it is best to place the top plate at the outside of the middle third of the wall thickness. It may also be worthwhile to choose raking stones on the outside face to accommodate the rafters, to save having to trim each rafter where the wall is too thick for it to clear the edge of the wall. A top plate of

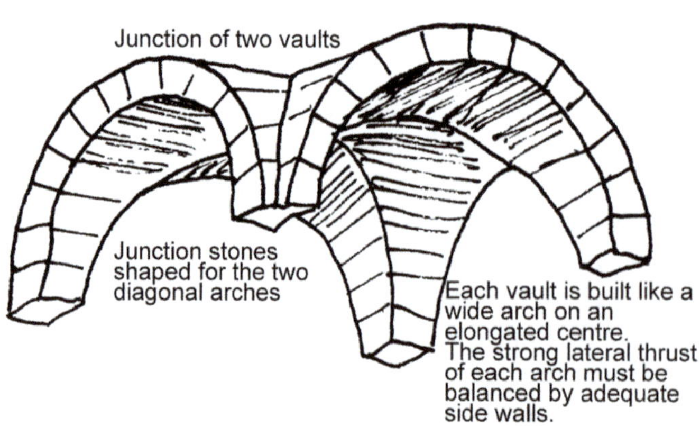

Figure 7.31: Vault.

Figure 7.32: Window in gable end over lean-to roof at rear, Bombala, New South Wales.

75 mm × 100 mm (3 in. × 4 in.) may be preferable to the conventional 50 mm × 100 mm (2 in. × 4 in.) for the extra height it gives at this location.

If used on the inside of the wall, face stonework looks best if the stone is built up to ceiling height, or to a cornice at the edge of the ceiling. To achieve this, the wall can be brought up to level where the top plate will sit, but left unfinished on the inside. After the top plate is in place the last course can be finished with small stones either up to the top of the top plate or higher in the case of a sloping ceiling.

Parapets and upper walls

In places where stonework is built higher than an adjacent roof, care must be taken to ensure that moisture cannot penetrate to the wall below the roof. It is particularly important that parapet walls do not absorb water. A course of coping stones shaped with a peak along the top of the parapet wall can shed water away from the wall, or the wall may be finished off with stones similar to those in the wall underneath, but with waterproof mortar used between them. Sometimes bricks on edge are used to cap stone parapets, or they may be rendered (Figure 7.33).

In addition to this, a damp-proof course and flashing should be placed immediately above the roofing material to stop wind-driven rain from penetrating the wall at this very exposed location. Built into the mortar below the DPC, the

Figure 7.33: Concrete windowsills and structural elements are shaped to give character to the basalt walls of this outbuilding, with a relieving arch over the doors, lintels expressed as straight pointed arches, and round openings for the roof vent and side windows, Bathurst, New South Wales.

Figure 7.34: A bat motif, carved into the base of the sandstone chimneystack at Abercrombie House. Framed above it is a plaque dated 1878. Smooth sandstone wraps around the corners to frame the windows.

metal flashing is bent down to the roof surface. To accommodate a roof slope, stepped flashings are built in for each course, each overlapping the one below. Because wind driven rainwater has the potential to collect, the edge of the roofing also needs a layer of flashing lipped up under the wall flashing.

Flashing is also necessary at changes of roof level and around chimneys (Figure 7.32). All the places in a house design where water may be able to work its way in during extreme conditions of flooding or driving rain should be anticipated, and measures taken accordingly.

Figure 8.1: Incidental seating at a change of level is helped by a large stone slab serving as a table, in a sunny location, near Hepburn Springs, Victoria.

'Rocks have always been called "living" in their native place.'

John Ruskin (1865), *Ethics of the Dust*

8

Stone in the landscape

In the natural landscape, rock formations highlight many of the most spectacular scenes. Rock plays an important part in the pattern of landforms, the shape of coastlines and the direction of rivers. When rock is used to punctuate our human environment, it can be used in a formal, contrived manner, or in a way that enhances its natural qualities. Sometimes the combination of the two can work successfully, such as the use of a rough rock face in a geometric form (Figure 8.8).

A rock may also play a symbolic role in the landscape. Smooth geometrically shaped rocks or carefully finished sculptures express the imprint of mankind among the randomness of nature. In time, a smooth stone weathers and grows a covering of moss and lichen over itself. In the same way, a statue may lose a nose or an arm, and become smaller, perhaps closer to the shape that originally inspired the sculptor to work the larger rock. Some rocks are naturally carved, by water or other forces, into forms that resemble living creatures, and their placement in a garden brings the place to life, giving character.

In the course of collecting stones for a house, some stones found can have an aesthetic quality that would be difficult to retain if they were built into a wall. It feels wrong to remove a beautiful covering of moss and lichen from a stone, even if it is a good shape to build with. Weathered, lichen-covered stones are usually brittle, so they cannot be easily shaped or trimmed. If the mossy surface is on the face of the stone it can be left there and the stone built into the outside of the wall. Moss or lichen on other surfaces is another matter. A mossy stone that fits well into a difficult part of the wall warrants cleaning off, but if a stone is irregular it may be worth keeping for the garden rather than trying to shape it.

Figure 8.2: Retaining wall built on rock outcrop, using excess stones from quarrying. This type of wall is also known by the Scots as a ha-ha, near Hepburn Springs.

Quarried stone is unlikely to be one hundred per cent suitable for easy wall construction unless the stones have been carefully selected. Where a structural stone wall is being built with level top and bottom surfaces along the courses, some stones will be rejected. Although larger stones often look as though they will break up neatly to a buildable size, a portion of them do not. In addition, a quantity of small stones are formed from the trimming or accidental shattering of other stones. As a result, the building site can end up being surrounded by stones that will not be used in the walls of the house.

Rejected house rocks such as these can be used for rough dry rubble walls, borders around garden beds, or retaining walls (Figure 8.2). They can be used for artistic effect or to facilitate height changes in a rock garden.

Quarry reclamation

When a quarry has been worked, there are usually a number of rocks lying around that weren't taken to the house site. Some may be good enough to build with, but were second choice and thus left for another load. Others show no promise of breaking into favourable pieces, or are oddly shaped broken fragments that were left behind when the good stones were taken out. These fragments, along with overburden from areas still to be worked, can be used to back-fill a worked-out quarry pit (Figures 8.3 and 8.4). Once filled, this waste is compacted, and the topsoil from a new quarry area can be put on top (or the original replaced).

Figure 8.3: Solid rock reef edges a lawn area levelled after quarrying was done. It was not worth breaking up this rock because there were no straight cracks to develop.

If the quarry is not likely to be used again, either because the accessible rock supply has dwindled, or because the good rocks have all been taken, it should be filled in and reclaimed. If the quarry was initially worked in a way that separated the topsoil, subsoil and reject rocks into different piles, it should be simple enough to push these back into the quarry pit in the correct order. The rocks are used to fill the low spots and establish landform, the subsoil to smooth out levels and to

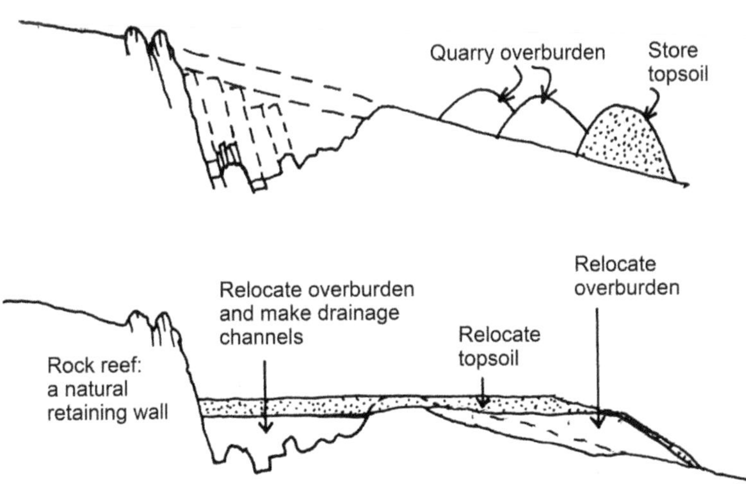

Figure 8.4: Quarry reclamation.

Figure 8.5: Local flagstones used for paving, including swimming pool steps, near Bairnsdale, Victoria.

make drainage slopes, and the topsoil to cover it, providing adequate depth for the roots of plants.

Depending on the environment, it may be necessary to do some seeding or planting as well. This can stop the intrusion of unwanted weeds – usually the first plants to establish themselves on used quarries, mine tailings, roads and other scars on the landscape.

The stones suitable for building should be grouped where they can be collected easily. They will be useful if the house is extended in future, or they can be given to other stone house builders.

Some used quarries lend themselves perfectly to water catchment and a water-collecting quarry is easily reclaimed as a pond. An old quarry that retains water will probably be surrounded by reeds and other water plants soon enough, but the slopes leading to it should be landscaped to prevent weeds establishing themselves there.

Water and rock

It may be a help in designing gardens for rocky locations to learn from traditional Chinese and Japanese garden philosophy. Adding water to the design can help to create a special place (Figures 8.5 and 8.6). The contrasting elements of water and rock are ideally balanced and are especially valued in Oriental gardens. They are the embodiment of the symbols yin and yang. The Chinese word for landscape, 'shan shui', means hills and water. Hills are the yang element, represented by rocks, which provide stimulation and are balanced by water, which induces tranquility.

A Japanese guide to garden construction from the Heian period discusses the direct relationship to good and bad luck[6] as well as other general advice on proper

Figure 8.6: Water channels make a feature of the irregular edges of paving stones in a path, Orbost, Victoria.

rock placement. For example, to set a stone, which was naturally flat, upright may invite bad fortune. Although they may sound like mere superstition, these rules of garden design are mostly derived from observations of nature and aesthetic considerations (Figure 8.7).

In a garden, a water feature brings to mind a stream or lake. As in nature, rocks should be placed where a stream curls around a hillock, or empties into a pond, or they should be where the stream enters a courtyard or curves as it passes a building. In addition, other stones can be sited for aesthetic effect. The bend of a stream is the most suitable location for rock placement, since it prevents the point of land from being washed away, as in nature.

Rocks are sometimes used to form an island or to represent a boat on the water. In an ornamental garden these rocks may be placed along a sight line from a building or a path, to enhance the view, or they may have some symbolic representation.

Symbolism

The European tradition of garden walls, pavilions and ornaments built in worked stone displays a formal geometry (Figures 8.9 and 8.10), sometimes based on classical temples. Sacred sites have had their elements rigorously copied to adorn gardens, as have aqueducts, mausoleums, fountains and triumphal arches of

Figure 8.7: Rock and water: design relationships.

Figure 8.8: Castlecrag gateway with crystalline gateposts in a rock face around a patterned garden gate.

Figure 8.9: Vines soften the concrete edges of this Castlecrag balcony.

Figure 8.10: High basalt garden wall and capped gatepost for courtyard to the homestead, Narrapumelap, Wickliffe, Victoria. The wall is an extension of the house wall, and serves to unify the house with the outbuildings.

Figure 8.11: Ornamental garden wall at Abercrombie House, Bathurst, New South Wales.

classical Greek or Roman design. As a result, in formal European style gardens, and in public parks with their paths, gateways, walls and bridges (Figure 8.12), the character reflects a value placed on rational thought and dominance over nature.

In a garden, the imposition of geometric order on the organic background of plant forms is symbolic of mankind in control of nature (Figure 8.11). Or it could be an artistic reference to the natural formation of crystals in the stone. It also represents wealth, since detailed stonework is time-consuming and expensive.

In contrast, the traditional Oriental ideal of a garden tries to evoke the feel of a simple mountain path. The path of the imagination might lead to an isolated monastery or temple, made from simple rustic materials, appearing natural, yet it

Figure 8.12: Classical European bridge.

Figure 8.13: Chinese bridge.

may still be contrived. The garden is conceived as 'built' from the ground up, instead of just 'planted' as it would be in the Western concept. The landscaping includes the creation of level changes and steps or paths leading between features (Figure 8.13). A carefully designed garden should ideally display the same balance of yin and yang as the forms of nature. The mountain and stream landscape is the most favoured, as the natural splendour of mountainous places is supposed to be uplifting for the spirit.

Oriental symbolism in gardens derives its history from early gardeners' intention to attract the good spirits who lived in the clouds. Rocks distorted by erosion into animalistic shapes were especially valued by the Chinese if the creature they resembled had a favourable symbolic meaning. The Chinese garden was planned as a series of views, using piles of rocks to give direction to pathways and to represent mountains, especially the mythical peaks of the Western Heaven. Open pavilions were placed where they would give the best view of another part of the garden: built with a lightweight structure so as not to weary the spirits, who were accustomed to the clouds. Walls might have openings to frame a view, such as a moon gate, where the circular form draws the focus to the centre.

Entrances to smaller Oriental gardens traditionally have an interior screen wall, for privacy, and to keep out the bad spirits, who can only travel in straight lines. Zigzag bridges have the same function (Figure 8.14).

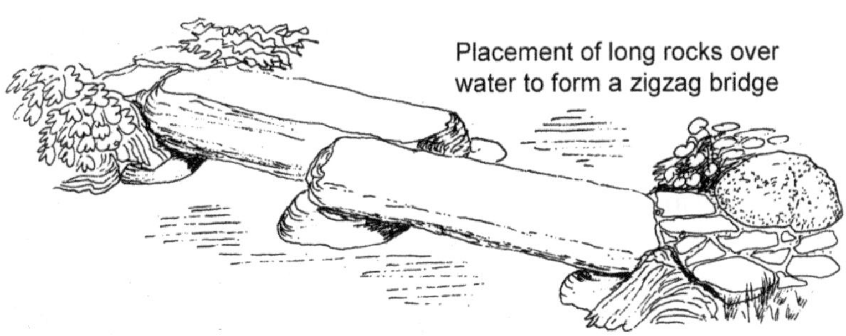

Figure 8.14: Zigzag bridge.

Figure 8.15: The barbecue is often a stone feature of the Aussie backyard, but in this Castlecrag example it has been taken over by plants.

A garden can be made to seem bigger by deliberately setting up rock configurations that mimic a wider landscape and accentuate its natural quality. Pathways that link between them can offer views from a distance. Pavilions for contemplation would be placed with a view of a rock feature, to emulate temples placed next to awe-inspiring rock formations. Appropriate in such a place would be a trinity formation of a large central stone flanked by a smaller stone on either side, or a pool with one or more islands.

In both Chinese and Japanese traditional gardens, rocks have been used to create islands representing the three mystic isles of the Eastern Sea. This was the home of the immortals who reputedly had palaces on high terraces. In a garden, such islands are surrounded by water, moss or raked sand (rippled to represent water), and include flat-topped rocks at different levels.

Another symbolic island form is composed of flat rocks located to represent a tortoise (a symbol of longevity). The horizontal form of the tortoise island can be contrasted with a crane island made from vertical rocks to represent a crane. The rocks used rely on the imagination of the viewers in their interpretation of the horizontal or vertical form, which is enough to create the symbolism. Stones shaped into picturesque forms by weather or by water were much valued for use in

Oriental gardens, and were at times transported from older gardens to enhance gardens under construction.

The stone barbecue

More commonly than sculpture in the Australian garden, the stone feature valued most is the stone barbecue (Figure 8.15). This is an accepted part of our culture, and if challenged, any stone collector need only say they are collecting rocks to build a barbecue, and they are usually welcome to continue! A barbecue is only a small project, requiring relatively few rocks, which should fit neatly into the boot of an ordinary car.

Garden spaces

External space is valuable around a house, especially if it is sheltered from the wind and is sunny in the early morning or late afternoon (Figure 8.16). So the layout of external space should not be an afterthought. The space created in a garden is important for how it can be used, and should shape the designed landscape. Plants are important in helping to provide shelter and an edge effect for the space. Rock features in a garden also help with this, especially where there is a change of level.

Coming and going from a house, it is beneficial to have spaces where people feel comfortable to stop and talk. These spaces can be informal in character, and a change of level can provide incidental seating, making a place to relax in and look around for a while, or to spend some time with a friend. Steps are often

Figure 8.16: Granite table made from a large rock, the centrepiece of a terrace where it was quarried, near Moruya, New South Wales.

Figure 8.17: Large stone steps leading down to terraced area, where stone was quarried, next to stone tower of house near Hepburn Springs.

used this way, as an extra seat, and so are garden walls, if the height is right (Figures 8.17 and 8.18).

Garden walls

The garden wall often serves the purpose of giving privacy, but in some locations it is needed more as a retaining wall. Stone is ideal for retaining walls that rely on their weight and solidity to prevent landslips. If the height exceeds about a metre,

Figure 8.18: Granite steps in grey and pink, combining found and quarried stone, near Moruya.

Figure 8.19: Sculptural slate wall done in traditional dry walling, Adelaide Botanic Gardens.

more thought should go into structural design, to ensure that the centre of gravity of any unstable soil pushes towards the base of the wall below the footing (Figures 8.20 and 8.21). A series of low walls can help with this, and provide a route for a path as well.

Figure 8.20: Culvert in wall of rubble granite, Tasmania.

Figure 8.21: Garden walls built over boulders at former quarry edge.

The landscape as structure

Using natural stone in the landscape as part of the structure of a house can be an interesting way of blending a totally different built form with its natural context. A number of houses have been built incorporating large stone formations as walls or protrusions into the house with good effect (Figure 8.22).

Caves and cliff dwellings have been lived in by people in the past and present all around the world. According to workability of the available stone and other resources, people have built their homes in contact with the living landscape using a wide range of techniques. In some locations, dwellings can virtually be carved out of the parent rock, or walls easily built from fallen rocks.

A dramatic stone cliff or outcrop may in itself create pleasant living spaces, cubbyholes, storage niches and fireplaces. Where impressive rock formations occur on private land, people naturally enjoy spending their time there, especially where the rock offers sheltered sunny places. Some such places may contain the remains of ancient fires. Some are already comfortable as outdoor spaces and would lose much of their charm if a building were placed in their midst.

Figure 8.22: Using the space between the living rock as a building site.

Building alongside a large stone feature is a challenge in working sensitively with nature, which should not be taken lightly. The context can create unusual engineering problems, or provide a setting for unorthodox design decisions. The house takes on a sculptural quality. Walls may be better with a slope, to work with the flow of the natural rock rather than against it. A curve or a cantilever may be desirable in another part of the building to give it sculptural balance. The juxtaposition of views with the rock should be assessed and retained as much as possible. Moss and lichen left on weathered rocks can help to retain the feel of a naturally formed environment. Care must be taken not to spoil the original quality of the natural rock formation.

Natural landscape

Throughout Australia, many Aboriginal sacred sites include rock formations representing symbols or creatures of the Dreamtime. Although the traditional culture of the Aboriginal people has been lost in many areas, the land can still be read for images that are evocative of this symbolism. An impressive rock formation is no less deserving of respect today, and should be left intact for its own sake so that it can be enjoyed in its natural state.

Whether for the view they offer or for their symbolic role, the rocks used in landscaping should be considered as part of a complete setting for the way they balance their natural surroundings of plants, water, sand or soil (Figure 8.23).

Figure 8.23: View of rubble house in its rocky setting, Yandoit, Victoria.

Figure 9.1: Farmhouse chimney near Bairnsdale, Victoria, with tuck pointed mortar joints.

'Air-intake control is the key to efficient fireplace construction. The ignition of a correctly proportioned gas-air mixture will give complete combustion of wood and emit gases containing only the non-combustible carbon dioxide, water, oxygen and nitrogen …'

'One of the more valuable experiments which evolved from the Prairie States Shelterbelt Project … a windbreak not only provides fuel to burn but fuel savings as well.'

Ken Kern (1974), *Producing Your Own Power*

9

Fireplaces and chimneys

A stone fireplace can create one of the most enjoyable rooms to relax in a house. The balance of a lively, flickering fire with the natural or sculptured texture of stone makes an ideal combination. The fireplace is best positioned so people sitting by the fire are in a sheltered part of the house, away from any draught from a poorly sealed door or window. A pipe bringing air from outside the house can supply oxygen to the fire, so that less fresh air from the room is drawn up the chimney.

The internal lining of a fireplace is not generally faced in stone, because very few stones are totally resistant to heat. In general, sandstone is not seriously injured by heat, but it quickly becomes stained by smoke. Limestone does not lose its strength when exposed to temperatures up to 450°C, and seldom cracks from the heat, but it may be reduced to powder under the application of water afterwards. Granite is not fire resistant and generally cracks as the various minerals in it expand at different rates. Basalt also is likely to splinter if in contact with fire. Although many old fireplaces were lined entirely with stonework, they were usually built to a depth where the heat of the fire would be slightly away from the stone, and they were not very efficient at throwing heat into the room.

To help draw the smoke, the traditional fireplace was deep with a wide flue. It was found that the depth could be lessened. The Rumford fireplace of the early 18th century was made with an opening as high as it was wide. For efficiency, it was built with less depth than most at the time, with a very shallow angle for the side walls of the firebox (Figure 9.2). The throat width of a Rumford fireplace is 100 mm and it uses no metal, in the belief that radiant heat from the masonry lining is superior.

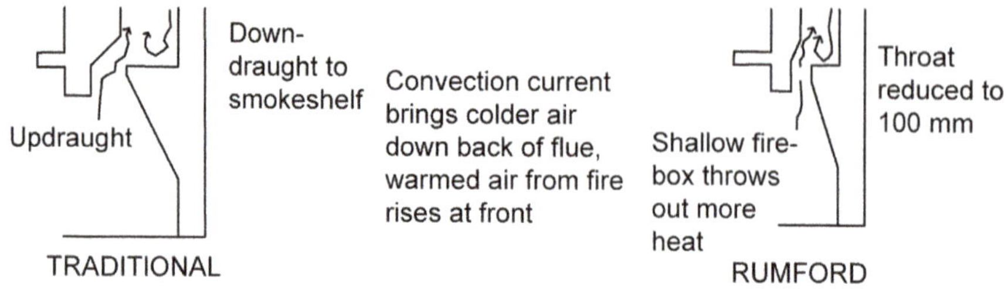

Figure 9.2: Firebox.

For shallower, more efficient fireplaces, it is best to use firebricks to line the firebox as they withstand the heat. Soapstone is also suitable, if it can be found. In some places it is available in slabs for firebox construction. A few other soft and homogeneous stones may be suitable to use – although in general, stone is unlikely to last as long as firebrick.

Any stone that is planned for use in contact with fire should be well tested first in a hot camp fire to establish that it will not explode or splinter. Rocks containing moisture are the most likely to explode when heated, as the water changes to steam, exerting pressure to split the stone. Firebricks are made from clay with a high alumina content, which gives them a smaller coefficient of expansion. This makes them less susceptible to thermal shock from the continuous heating and cooling of the fireplace. Failing the use of firebricks, a traditional method of protecting the masonry of a firebox is to periodically render it with pipeclay.

The easiest to build with are firebricks the size and shape of conventional fired clay bricks. Narrower but higher, square firebricks are also available. These are made for lining wood-burning stoves and can also be used to build the firebox of a fireplace. However, they are thin, and in time they are more likely to tip out of place. Refractory cement or fireclay should be used as mortar with firebricks where the fire will be in close contact.

Proportions

The shape of the firebox should lean inward to throw the maximum amount of heat into the room, yet send all the smoke up the chimney (Figure 9.3). If the firebox is too deep, much of the heat generated is lost. The sides of the firebox can be built at an angle so that heat is reflected and radiated into the room. The upper section of the back is built leaning forward to direct heat towards the lower parts of the room. At the top of this leaning wall, a space is created at the base of the chimney for the smoke shelf. The smoke shelf stops any downdraught from disturbing the upward flow of smoke from the firebox.

Figure 9.3: Flat arch shaped in rock-faced stone over fireplace, Bungendore, New South Wales.

The size and height of the chimney flue will also have an effect on how well the fireplace draws. If the height or cross-section area of the flue is too small, then smoke is more likely to enter the room. A fire usually draws better once established, with convection currents in the flue responding to the heat. If a chimney sits among taller buildings, wind blowing towards them may increase downward pressure in the flue, and its height should be increased.

Table 9.1 shows proportions for the elements in fireplaces, indicated below (Figure 9.4). The chart is a guide to the efficient design of a fireplace, for a range of room sizes.

If a separate branch chimney is provided for a fireplace, the design may change. In the alternative arrangement, a continuous main chimney rises from basement to roof. Each fireplace is connected with its flue at an angle, and any ash or soot build-up falls to below where there is a clean-out chamber, with a well-sealed door.

The walls of the firebox should be laid with firebricks in refractory cement at the angles needed for the fireplace to perform well. Dimensions must be worked out before construction begins. A plywood template helps to align the bricks where the fire back tilts forward.

The fireplace opening should be from 8 to 15 times the area of the flue, depending on the flue height. If the size of the fireplace is too big for the room, it draws too much oxygen from the room, and poor combustion may be the result. The actual shape of the fireplace opening is usually built wider than it is high.

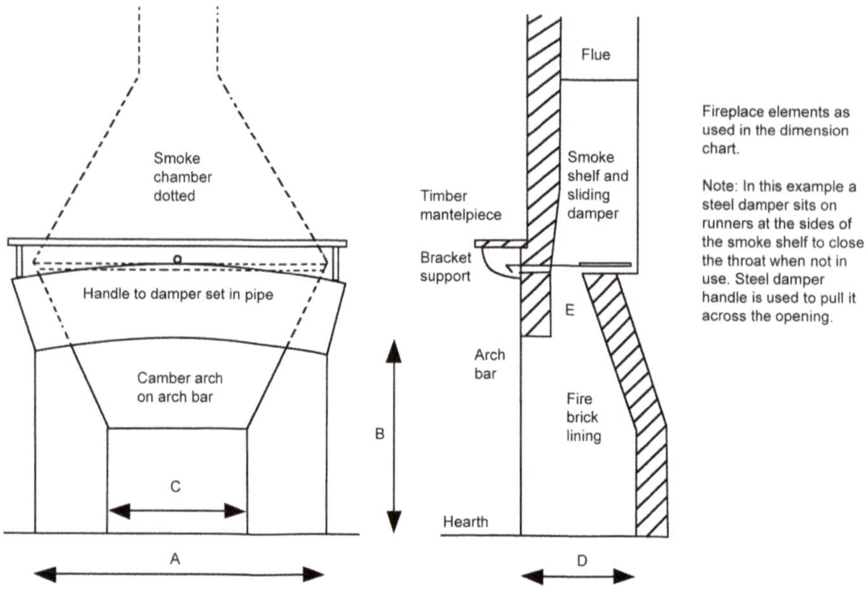

Figure 9.4: Fireplace elements as used in dimension chart (Table 9.1).

Figure 9.5: Granite and brick fireplace, with steel doors fitted to reduce air intake, Dandenong Ranges, Victoria.

Figure 9.6: Back of fireplace, large granite blocks roughly squared at corners.

Table 9.1: Suggested fireplace dimensions

Room size		Fireplace						Flue (minimum dimensions for height)			
Area, sq m	Volume, cu m	Width A, mm	Height B, mm	Opening area, sq m	Back width C, mm	Depth D, mm	Throat E, mm	Up to 4 m FxG, mm	Up to 8 m FxG, mm	Up to 12 m FxG, mm	Up to 16 m FxG, mm
16-22	40-60	600	500	0.30	360	340	100	250x250	200x200	200x200	200x200
		650	550	0.36	400	350	100				
22-30	60-90	700	580	0.41	440	360	100	250x250	200x200	200x200	200x200
		750	600	0.45	490	370	100				
		800	630	0.50	530	380	120	250x370	250x250	200x250	
30-40	90-120	850	660	0.56	580	380	120	250x370	250x250	200x250	200x200
		900	680	0.61	620	400	120				
		950	710	0.68	660	400	120				
40-50	120-180	1000	740	0.74	700	420	120	370x370	250x370	250x250	200x200
		1050	760	0.80	740	420	120				
		1100	780	0.86	780	450	120				
50-70	180-250	1150	820	0.94	820	450	150	370x370	250x370	250x250	200x250
		1200	840	1.01	850	480	150				
		1250	870	1.09	890	480	150				
70-90	250-350	1300	900	1.17	930	510	150	370x490	370x370	250x370	250x250
		1350	920	1.24	970	530	150				
>90	>350	1400	950	1.33	1000	540	150	370x490	370x370	370x370	250x250
		1450	970	1.41	1050	550	150				
		1500	1000	1.50	1090	580	150				

Fireplaces with arched openings may benefit from a higher opening in the centre than that listed in the chart (Figures 9.5 and 9.6).

Dampers

For increased efficiency, a damper can be installed across the throat so the fireplace can be closed off from outside air when not in use, and it can also be adjusted to reduce the heat loss when the fire is burning low. The damper should be removable, so that the smoke shelf can be cleaned periodically, or in case replacement is necessary. Cast iron dampers are less likely to corrode than steel plate.

It is possible to buy a damper of standard construction. Otherwise a damper must be made up especially for the fireplace. The dimensions of a standard damper will influence the construction. It should be chosen in advance, so the fireplace can be built to suit.

Construction

As each course of the back wall is completed, the same course of the sides should be laid alongside. The stonework of the walls beside the fireplace is built in front of, or beside, the firebox lining, so enough space should remain to fit the stones chosen to go there.

The throat runs across the width of the fireplace, 200 mm to 300 mm above the lintel or arch of the opening, with a gap of 100 mm or more for the smoke to go through.

The smoke chamber above the throat extends across the whole width of the fireplace, with its sides sloped at an angle from 45° to 60° up to the chimney flue. The walls should be smooth so the smoke is not slowed down. Standard bricks or a flue liner are better than stone to give a smooth finish. The flue is best placed above the centre of the fireplace. The first 500 mm of flue above the smoke chamber should be vertical. To ensure easy cleaning and to avoid eddy currents in the smoke, an equal flue area is necessary from top to bottom.

Fuel-efficient fireplace designs include the use of metal liners or convection pipes. The advantage of a steel firebox is that air from the room can be heated by convection behind it and returned to the room. Steel pipes can serve the same purpose in a masonry fireplace, running from the hearth level, under the fire and over the top.

Pipes can be taken from a room or passage behind the fireplace and over the top of the fire, bringing extra heated air into the room (Figures 9.7 and 9.8). Hot water pipes can be fixed behind the plate steel of the firebox as an auxiliary to the hot water system. This is commonly known as a wetback.

Some refinements to the design can make cleaning easier, but are rarely used. The smoke shelf area is difficult to clean without bringing soot and fly-ash into the

Figure 9.7: Narrow firebrick lining, with slots for insertion of steel convection pipes and arch built of recycled church stone, near Hepburn Springs, Victoria.

Figure 9.8: Mantelpiece being propped while mortar sets.

Figure 9.9: Asymmetrical stone mantelpiece and side shelves, with chimney under construction. A clay model was used to resolve the design, determined to suit the mantel stone, near Hepburn Springs.

room. If the back of the fireplace is in a corridor, a service area or outside, a cleanout door can be located there, and it is especially useful if the interior of the room is plastered or wallpapered. The cleanout door should be fireproof and well sealed.

An ash trap and ash dump below the floor may be provided at the base of the firebox, so that ash and charred wood can be cleaned out from below, either in a basement or outside at a lower level (where the fireplace is in an exterior wall). The ash dump must be set in fireproof material and be easily accessible.

Where metal is built into the masonry of the fireplace, allowance should be made for its greater rate of expansion when exposed to high temperatures. Steel is one of the most satisfactory materials for this use, although it is susceptible to rust,

Figure 9.10: Smooth round ledges on pilasters either side, instead of a mantelpiece above, for this fireplace with a gas heater installed, Castlecrag, New South Wales.

Figure 9.11: Chimney incorporated into parapet at change of roof height, Castlecrag, New South Wales.

particularly from rainwater coming down the chimney, or if it is in direct contact with the flames. Cast iron may be preferable to steel plate, although it can crack under sudden temperature change. Stainless steel is harder to work with and the extra cost is seldom justified. Galvanised steel should not be used, as zinc produces poisonous fumes when burnt.

A number of manufactured fireplace inserts are available. These have duct openings that must be allowed for in the stonework above the fireplace opening. Some of these units include provision for hot water pipes that can be incorporated into the hot water system.

Figure 9.12: Fireplace inside curved walls, contrasting smooth face figured stone with rock face of side ledges, Castlecrag.

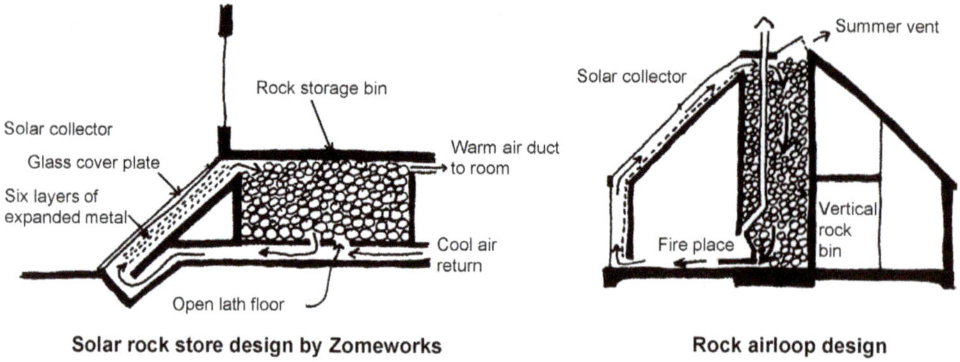

Figure 9.13: Rock heat storage applications.

They may also provide ducted heat to other locations. Building a ducting system of this type behind the firebricks, in the stonework, would not be as satisfactory for heat exchange because stone does not heat up as quickly as metal.

The hearth of the fireplace should not slope away from the grate. It should be built of masonry to be fire resistant, ideally at least 75 mm above the floor, and extend a minimum of 350 mm beyond the front and 150 mm beyond the sides of the fireplace opening. Alternatively, the hearth may be contained within the fireplace opening, in which case it should be a minimum of 155 mm above the floor. A masonry or tiled area of floor should be laid adjacent to this, extending to the same dimensions as for a raised hearth.

Mantelpieces are conventionally built at a height of 1200 mm above floor level. For cantilevered stone mantelpieces, this involves a fair degree of precision and forward thinking. In particular, it requires the right stone, which must be propped while the next few courses of stonework are laid and set in place, to hold down the part of the stone embedded in the wall (Figure 9.9). An alternative is to place a ledge on either side of the fireplace to achieve the same purpose (Figures 9.10 and 9.12).

Storing heated air in a rock bed

Rocks can also be used for storage of heated air via ducting. Heat can be generated either by a fireplace or with hot air from a solar collector, or by both. If the rock heat storage, in the form of rounded pebbles (such as river pebbles), is placed higher than the air intake at the bottom of the collectors, the air can circulate by thermosyphon (hot air rises), thus avoiding the need for pumps. The heat storage capacity is a result of the long storage retention period that is a natural characteristic of the rock (Figure 9.13). Adequate insulation and a compact shaped volume are essential to storing heat for the maximum time before pumping back when needed. (For more information, refer to Anderson 1977.)

Figure 9.14: The narrowing of the chimney above the smoke chamber is expressed externally at the gable end of this house, Hunters Hill, New South Wales.

Chimneys

The chimney should not be used to support structural timber unless there is at least 200 mm between the chimney flue and the timber. Metal flues may be used if necessary. They must be located with due protection for combustible materials and at distances stipulated by the building regulations. Metal flues will not last as long as masonry flues. They must be positioned in a way that will allow future replacement, preferably where they can be seen, so that rusting is obvious. Stainless steel is preferable as it lasts longer, but it too will eventually need replacement.

Where the chimney emerges from the roof, water must not leak through (Figures 9.11 and 9.14). A DPC and metal flashings should be laid in the courses directly above the roof. Stepped flashing is used to follow the slope of the roof. For a better fit, brickwork is often used instead of stonework for construction of the part of the chimney directly under the flashing.

The roof flashing is placed first around the base of the chimney, and on the slope it is overlapped between layers of roof covering. It is bent upwards around the sides of the chimney in readiness for the stepped flashing to cover it. The stepped flashing and other flashing strips are built at least 50 mm into the stonework with a DPC, and the lower edge is trimmed where it meets the roof slope. Each piece of flashing should overlap the next by at least 100 mm.

Where the chimney is at the lower part of the roof slope, the volume of water may be difficult to manage unless a gutter or a miniature gable (also called a cricket) is built on the upward side of the chimney, to shed roof water on either side.

There are a few rules of thumb, some of which were incorporated into the building regulations when it was usual for houses to have chimneys. The chimney flue should rise at least 300 mm above any roof within 3.6 metres of it. The height of the chimney above the place where it meets the roof should not be more than six times its least width.

With turbulence or pressure due to wind blowing towards an adjacent building, the fire may not draw properly until the chimney warms up. A higher chimney draws better and raises the flue above the influence of these effects. The top surface of the chimney can be sloped, or stepped down from the centre, so smoke blows away without eddying at the chimney top.

A chimney cap can serve to protect the chimney from rainwater, as well as stopping possible downdraughts. The design of the chimney cap should not obstruct the flue from being properly cleaned out.

A tunnel-shaped chimney cap can be aligned with prevailing winds to encourage a directional draught, and improve the draw of the chimney, especially

Figure 9.15: Shelves to the side of unbalanced mantelpiece, house near Hepburn Springs, Victoria.

if there is a chance of wind pressure increase from roofs or buildings nearby. The size of this tunnel should be as big as the flue but not much bigger. It should be parallel with the closest side of the higher building element. This encourages wind flow through the top of the chimney so it is unaffected by air pressure build-up from wind buffeting around the building.

Once the functional aspects of the chimney operations are sorted out, there is still the form of the fireplace to consider (Figure 9.15). It is a prominent feature in the room, so decisions about whether there is a mantelpiece, and how it presents itself, or if there are ledges, and where, all make a difference to the mood at the hearth, which is central to the house.

Figure 10.1: Stone paving in an old-fashioned bathroom, Yandoit, Victoria.

'Stone possesses great advantage over wood for stairs: First, it will not burn; second, it is more substantial both in appearance and reality; third, it is more enduring and stands less in need of repair; fourth, it requires no paint; fifth, stone stairs are better looking, and if to this may be added that they cost less to make, the reasons for their use would seem conclusive.'

Ernest Flagg (1922), *Small Houses, their Economic Design and Construction*

10

Floors and steps

Floors

Stones for flooring need not be regular sized or shaped as long as they have a flat top surface. It is easier to lay a floor with rectangular or hexagonal paving stones, and the finished effect may be preferable, but irregular stones, as always, can be collected much more cheaply and, if flat enough, can be laid to a smooth finish. The mortar infill between irregular stones needs to be wider to accommodate the irregularities, and the bed where they are laid must allow the deeper stones to be set lower than the thin stones.

The stones should be laid on a solid surface, the ground or a concrete slab. They should not be laid on timber flooring unless the stone is thin like tiles and the floor has been adequately prepared and well waterproofed. As timber movement can cause cracking, wire mesh should be used beneath the stone if it is laid on a timber-framed floor. As with tiled floors, it is usual to have a fibrocement underlay. Stones of 80 mm thickness or less are the most convenient to work with.

Thicker stones can be laid directly on a sand or gravel bed with a layer of building plastic for waterproofing. If some stones are thicker than others they must be laid deeper into the bed. The stones are fitted into place first to ensure that they will sit at the right level. Then they are moved aside and the mortar bed is trowelled on. Very thin stones are better laid on top of a concrete slab.

Mortar for a stone floor should be stiff. The stone should push the mortar down as it fits into it. Then it is tapped to the desired level (Figure 10.2). If the stone

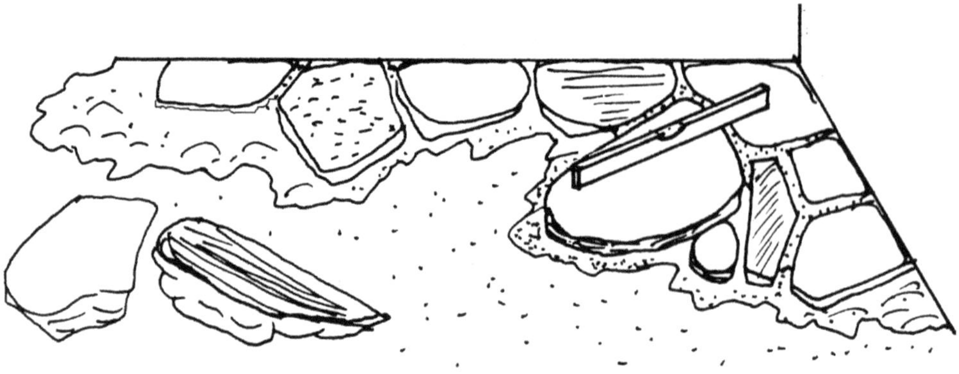

Figure 10.2: Laying floor from wall edge.

sounds hollow when tapped, then it is probably not seated firmly in the mortar bed, and may loosen or crack later.

The corners and the edges of the floor should be laid first. The level can be established with string lines or with a spirit level and a straight-edged stick (at least 1.5 metres long) (Figure 10.3). Once the edges are laid, the next stones should be laid level with these, gradually working towards the centre of the floor. The joints between stones may be filled with mortar as each stone is laid, or they may be filled afterwards. The stones should be dampened before the mortar is applied between them. The mortar finish is best if it is level with the top of the stones.

Figure 10.3: A piece of timber being used as a straight edge for setting paving into place, a bucket of water and brushes being kept handy to clean off each stone when in place.

A strong and fairly waterproof mortar mix for a floor is 1 part cement to 2.5 or 3 parts good graded sand.

In wet areas such as the bathroom or laundry, the floor is usually sloped very gradually (about 1 in 100) down to a drain. All drain pipes and plumbing should be in place before any work is done to pour concrete or lay stone for the floor.

Stone hearths and sills may be laid in the same way. Where a hearth is laid next to a timber floor, it can be built to cantilever from the fireplace construction or else constructed to full thickness from ground level. A cantilever can be built in reinforced concrete, then paved with stone, or in structural stone by corbelling, with each course of stone extending further than the course below until the correct height and width are reached.

Sills for doors or windows are laid with a slope towards the outside so water cannot collect on them. Waterproof mortar should be used between the stones.

Outdoor stone paving need not use mortar or a waterproof membrane. Stone paving laid without mortar may take a while to settle into place with the weight of people walking over it. The stones are fitted as close together as possible, and then sand, pebbles, clay or a mixture of these is brushed between them to fill in the cracks. If moss or grass between the stones is not wanted for any reason, cement can be mixed into the sand to fill the cracks, and the mix will set harder when the paving is wet.

Figure 10.4: Natural stone pavers, Yandoit, Victoria. While there is some roughness in the individual stones, care can be taken to bring stones to level, and better stones can be located where they are most noticed.

Suitable stones for floors or steps are slate or other layered, smooth-surfaced stones (Figure 10.4). Granite or marble off-cuts may be found in quarries or sometimes acquired from stone merchants cheaper than larger stones. If stones of varying colour or shape are used, a geometric arrangement or a pattern in the floor can help to give it a better sense of regularity. Otherwise the colour change looks like an afterthought.

Heating and insulation

Concrete or stone floors can have heating pipes set into them as they are built. Some concrete slabs have lines of electric element set in them for electric heating. This gives the householder no flexibility, as only electricity can then be used for slab heating, and operating costs are unlikely to go down. Hydronic systems, using water pipes, are more versatile, as the water in them can be heated by electricity, gas, wood or even solar collectors (if temperatures above 50°C are maintained).

Hydronic heating pipes can also be laid into a solid stone floor (although this creates extra work in fitting the stones in place if they are of varying depth).

Insulation is important under a heated slab and around the edges. Expanded polystyrene (such as that from used fruit boxes) is adequate for this, although extruded polystyrene is better, as it works better under compression and lasts longer. It is wise to insulate the slab edges, especially in cold areas, for at least 1 metre of perimeter width. Underneath the centre of the slab, insulation is not as critical unless the floor is heated, since the centre will not lose heat to the ground very quickly.

A heated floor takes time to warm up, and to cool down. When a house is lived in all the time, a hydronically heated floor is ideal for slab or stone floors, which are otherwise cold. However, another form of heating is necessary, in the short term, after any absence from the house, and hydronic heating is not suitable for a holiday house or occasional retreat for this reason.

If a concrete or stone floor is located to receive direct solar heat through windows when the sun shines, floor heating is redundant except for prolonged cloudy periods. Under-floor insulation is of benefit in this situation, just as it is where the floor is heated in other ways, especially to slow down evening heat loss.

Steps

Stone steps are made from single stones, used as found or cut to size (Figure 10.5), or from smaller stones fitted together.

Rough stone steps must be carefully constructed, especially when smaller stones are used, or if there is a lot of irregularity of stone sizes. The stones must be matched so that the steps will have equal height risers and equal width treads.

Figure 10.5: Cantilevered steps, Sydney.

The steps are built from the bottom upwards. The tread width of stones laid for each step must have extra depth to accommodate the riser stones of the next step. A base may also be built from the bottom up for subsequent steps, as required.

A large quantity of square-edged stone is needed for steps (Figure 10.6). The treads should have straight edges and, if the stones are not perfectly flat, the tread should slope slightly outwards, to avoid puddles collecting. The stones on the outside corners of the treads should be lodged well in place so they will not knock out.

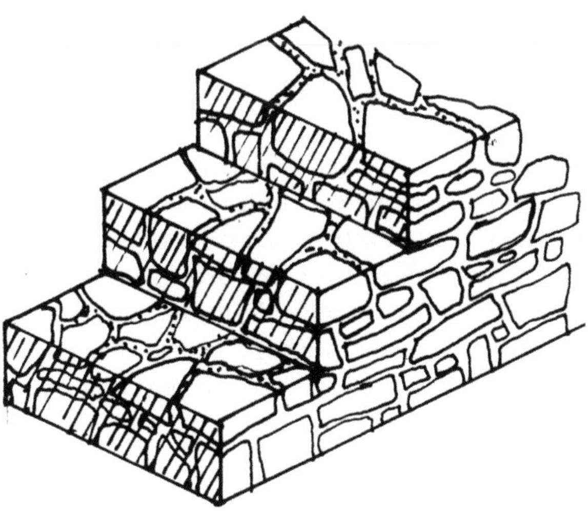

Figure 10.6: Rubble steps.

Figure 10.7: Steps and landing built of large slabs of stone.

Stone steps can be dangerous if not built carefully enough. To make a long flight of steps, a supporting structure is necessary for the upper steps (Figure 10.7). If the stones are small, this structure must either be solid stone or be built on top of a stone vault or concrete structure.

Stones that are long enough can be used to make steps that span across from one wall to another. The stones must be strong and of uniform texture, without cracks, to make good steps. Arches or lintels in one of the walls can stop the area below from being too enclosed. Stone steps can cantilever from a wall if there is enough extra length to set the end deep into the wall (Figures 10.8 and 10.9). Wrought stone steps or stairs are shaped from stone of good quality. Stairs of this kind made from solid stone steps are very unusual in houses, but may be practical occasionally when a house is sited with a local stone of very good quality. The degree of precision used depends on the whim of the stonemason. Traditionally, stone stairs were cantilevered from stone walls for stairways in grander buildings. Each step is rebated where it connects with the step below. The underside, or soffit, of each stone can be sloped at an angle to meet the next (Figure 10.10). The end of the stone that fits into the wall is squared, giving it a horizontal base to be set in the wall masonry.

Curved or geometric stairs require winder steps, wedge-shaped stones that radiate from the same centre point as the curve of the wall (Figure 10.10). A spiral

Figure 10.8: Curved cantilevered steps, in sandstone, leading to flat roof, Castlecrag, New South Wales.

Figure 10.9: Detail of curved cantilever steps.

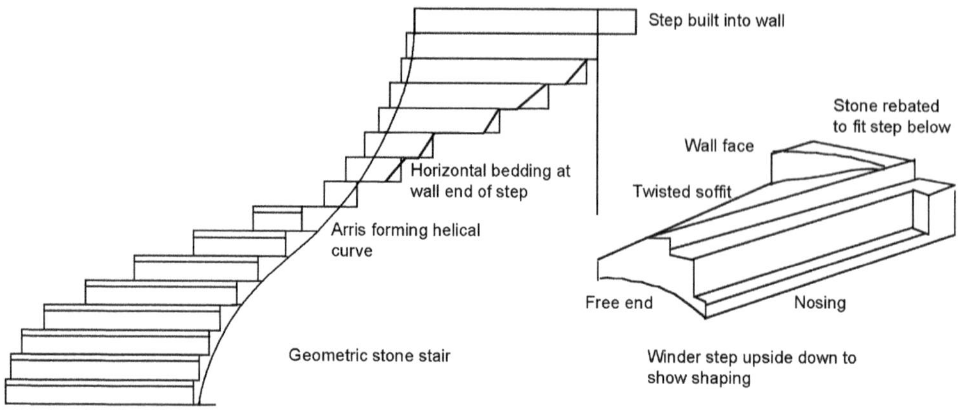

Figure 10.10: Geometrical stair and winder step.

stair can be built up with the cylindrical end of each stone stacking up to make a central newel post. This is a classic way of building a turret stair that has become less common now, as concrete stairs are used instead (Figure 10.11). The construction of geometric stone stairs is sound, but if repairs are needed the stones are difficult to replace.

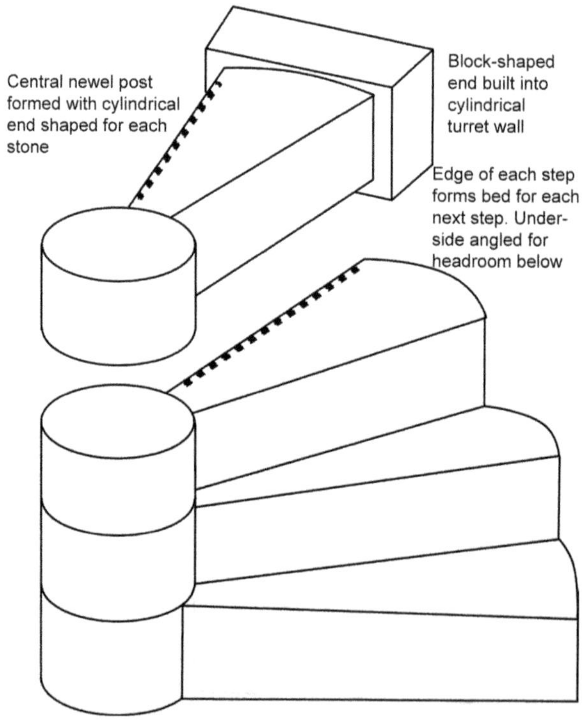

Figure 10.11: Turret stair construction.

Figure 10.12: Historic steps hewn in bedrock, The Rocks, Sydney.

Concrete steps faced in stone are poured in situ, to dimensions smaller than required: the size is reduced by the thickness of the facing stone plus the mortar that will set it in place. Any kind of stone facings can then be used for risers and treads.

Alternatively, steps can be formed as for concrete, but with stones for the risers laid in place within the forms. The mortar between these stones can be made more like concrete with concrete and rubble stone filled in behind them. The stones used for the treads must be placed carefully and should be of sufficient size to avoid dislodging at the edges. Reinforcing bars can be used. The formwork should be removed within a couple of days so the mortar can be cleaned from the surface of the stones.

Figure 11.1: Two-storey bay window, Abercrombie House, Bathurst, New South Wales, rock-faced granite walls with smooth-finished sandstone for corners and decorative elements.

'She is up and seizes plans and paper, and, before six o'clock, has enlarged the parlor very cleverly by throwing out a bow-window. So waxes and wanes the prospective house, innocently battered down and rebuilt with India-rubber and black-lead. Doors are cut out to-night and walled up to-morrow; windows knocked out here and put in there, as some observer suggests possibilities of too much or too little draught.'

Harriet Beecher Stowe (1868)

11

Drawings and specifications

Sketch plans

Once the house layout and room sizes have been decided upon, a sketch plan should be drawn. This is a scaled drawing showing wall thicknesses and allowing for measured positioning of such fittings as benches, sinks, cupboards, bath and so on. The intended furniture and furnishings should be measured so room sizes can be designed to provide adequate space for them. Ideas and concepts of how spaces will be used, and how the building will look, need to be weighed up with the initial structural information to ensure the design can be built.

Decisions should be made at this stage about building materials: which walls are to be in stone, timber, brick or mudbrick (as these affect the wall thickness and thus room size), how the floor is to be built and what kind of roof the house will have. If the land has a slope, the levels should be worked out too. As the job progresses, a number of changes are likely to be made, bringing the planning to the design development stage.

At this point, an assessment of energy efficiency can be done. Computer simulation using house energy rating software can predict energy use for a thermostatically controlled temperature range. It can also simulate the living and bedroom spaces to assess the likely range of summer and winter temperatures with no heating or cooling. This information is useful for working out any changes to window sizes and insulation before the design is finalised. Stone walls may not be available as an option in the number crunch by the computer, but mudbrick performs in a similar way (for the purposes of energy simulation). To achieve a

Figure 11.2: North side of stone house near Sunbury, Victoria. Pergola lets sunlight into living rooms on left, verandah is roofed on right for bedrooms.

better energy rating and reduce the need for summer air conditioning, many conventional houses use a concrete slab floor. Stone walls add enough thermal mass to perform well with a timber floor, as long as the orientation is right (Figure 11.2).

The sketch plans can be used for rough costing, based on floor area. Quantities of the different materials required can be approximately worked out at design development stage, to establish whether the proposed house will be the right size for the available funds. Conservative costing can balance out the smaller cost items that may have been forgotten. If costs are too high, plans can be altered easily enough in the design stages. Builders often estimate the rough cost of a job based on design drawings, but for a fixed price contract, more information must be given.

Working drawings

The local shire council will require copies of house plans and elevations, showing site and construction information for footings, walls and roof. The drawings submitted to council need to show that structural requirements are met, and are used as working drawings during the construction phase. Working drawings are part of the contract documentation for the building of a house, along with the written specifications (sometimes all the specification notes are written on the drawings). When a builder is employed or subcontract labour is used, additional detail drawings may be necessary for any work that is unconventional.

To create working drawings, a good knowledge of the building regulations and some skill in mechanical drawing is needed. Information showing compliance with the building regulations should be written on the drawing, as well as sizes of structural members and dimensions of rooms, wall thicknesses and relevant heights.

Working drawings are usually done professionally, although the owner builder can do them if necessary, as long as all the relevant information is sought out and written on the drawing. Footing design depends on the foundation soil, which may need a soil test. The council may be able to help in determining the soil classification and the appropriate footing design, but if it is not simple an engineer should be consulted. Charts are available to determine timber-framing sizes over most commonly used spans. If standard roof trusses are made off site, the engineering computations are done by the manufacturer.

Depending on the planning zone, the local council may determine that further reports are also needed to satisfy their requirements for fire safety, drainage, and retention of flora and fauna habitat.

Essentially, the construction drawings (working drawings) need to show the finished construction as it will appear when built (Figure 11.3). The method of achieving this result is left open to interpretation at this stage. The less detail

Figure 11.3: Curved bar fitted into a round stone enclosure, Bacchus Marsh.

shown, the more opportunity the builder can find to interpret the drawings to suit whatever building method for the details is currently economical. To achieve specific details and finishes, drawings at a larger scale, and specifications of particular requirements and products, can indicate what is required.

A usual set of drawings (to scale) for residential construction includes:

- site plan showing site location, house and distances from boundaries (1:200 or 1:500)
- floor plan or plans, showing required set out dimensions (1:100)
- external wall elevations (minimum of 3) and sections showing relevant heights (1:100)
- footing design drawing, or data sheet for conventional layout (various scales).

Further contract drawings may include:

- detailed floor plan, if required, in case of complex design or dimensions (1:50)
- cross-sections showing layout of critical construction details (1:50 or 1:20)
- internal wall elevations showing built-in items such as joinery (1:50 or 1:20)
- door and window schedule, indicating glazing layout (1:50 or 1:20)
- plans, elevations, sections of joinery, fixtures or other special items (1:20 or 1:10)
- details for structural connections, waterproofing, joinery etc. (usually 1:10 or 1:5).

Figure 11.4: The curve of the rock-faced façade is echoed by the smooth curved forms around the fireplace (see Figure 9.10), Castlecrag, New South Wales.

Specifications

The council requires copies of specifications. This can be waived for very simple buildings if the necessary information is written on the plans. Standard house specifications are available from some banks for brick veneer construction, which specify the basic requirements for all trades. These are adequate in most cases where the stonework requirements are written onto the plans. Anything that is not relevant to the particular house may be deleted, and specific finishes are filled in where required.

A more detailed description of the stonework may be needed in the specification, depending on who is doing the work, to cover the contract agreement. This is especially the case if techniques other than rubble stone will be used, to ensure a particular quality of work from the building contractor (Figure 11.4). For costing purposes, it is helpful to specify details of stone to be supplied and shaped by a quarry or stone merchant. For low cost rubble stone, only the basic details need to be specified (Figure 11.5). The supply of stone, the mortar mix, and cleaning down are the main thing.

In many cases the most expedient means of laying the stone is probably suitable for both structure and cost. What is shown in the drawings need not be described further in the specifications. It is usual to have a clause in the general section of the specification to cover any discrepancies.

Correlation is important for the construction elements necessary to achieve appropriate fire resistance or energy efficiency levels, as required for planning approval, including insulation, shading and window protection.

It is helpful if the specifications are concise and to the point, without any redundant information that does not apply to the particular job. The following notes are a guide to the kind of information that can be included in a specification (many of these clauses may not be relevant to the particular job being specified).

Figure 11.5: Knowing which parts of the wall will be seen, the stoneworker can locate individual rocks to their best advantage at eye level, house near Hepburn Springs, Victoria.

Figure 11.6: Fireplace under construction in basalt, Penshurst, Victoria, with decorative recycled stone mantel.

Specification notes: stoneworker

1. Selection of material: The quarry shall match the range of colour, texture and quality of samples already supplied and approved by the architect.
2. Drawings: The supplier shall provide drawings prepared from the architect's and engineer's details, showing all dimensions, key numbers and fixings, all of which should be checked and approved by the architect and contractor before any stone is worked.
3. Finishes: Exposed surfaces are to be finished in accordance with agreed samples (sawn, rubbed or tooled for ashlar, rough or chiselled surfaces for rubble).
4. Thicknesses and tolerances: Refer to thicknesses shown on drawings. The thickness of rubbed or sawn slabs should be within 3 mm of specified thickness, split-face slabs can vary up to 9 mm (appropriate to the stone type). As there are so many types of stone, the advice of the supplier should be followed (especially if it is in writing).
   For sandstone or limestone in vertical facings a minimum of 75 mm is recommended. For paving or steps, 40 mm or 50 mm thickness is permissible.

Marble wall linings should not be less than 20 mm thick internally and 30 mm externally.

5. Fixings: Metal anchorages for ashlar or facings should be made from a suitable non-ferrous metal. They should be of such shape and dimensions to be adequate to carry the loads to be imposed upon them. Care should be taken to ensure that all metal anchorages are properly positioned and secure.
6. Mortar: Masonry cement should comply with relevant standards. Lime mortars that do not contain cement should not be used in situations where high early strength is required.
   Typical mixes are:
   Lime:sand 2:5 (for limestone)
   Cement:lime:sand 1:3:12
   Cement:lime:sand 1:1:6 (for sandstone)
   Cement:sand 1:4
   Cement:sand 1:3 (for granite)
   Cement:lime:sand 1:2:6 (for marble flooring)
   The colour and porosity of the stone shall be matched by the mortar as closely as possible. Work should be kept free from mortar droppings at all times.
7. Damp-proof courses (DPC): These shall comply with the relevant standards and should be bedded in mortar not weaker than 1:1:4 (cement:lime:sand). Slates used for DPC should be in two courses.
8. Beds and joints: All stones are to be set on a full bed of mortar and tapped home, with all joints and joggles completely filled, and all mortices and cramps properly grouted.
   For rubble walling, beds and joints should be from 12 mm to 20 mm wide, depending on the regularity of the stone. For ashlar, the thickness should be from 4 mm to 6 mm, or 2 mm for carved or highly decorative features.
   The work shall proceed by courses, so that at no time shall new work be laid above previous work before the initial set of the previous mortar occurs.
   For large areas of cladding, compression beds shall occur once in each storey height, below the supported course. Expansion joints shall be provided at intervals of 6 to 9 metres.
9. Services: Adequate allowance shall be made for the inclusion of plumbing and electrical lines and fittings as required within the stonework. These are to be positioned as determined by the architect.
10. Scaffolding: Where a scaffold is required, it should be stable and safe for the imposed loads. Guard rails are required where the platform height exceeds 2 metres. All scaffolds must comply with statutory regulations.
11. Storage: Ashlar stones shall be stored clear of the ground and kept covered in winter. Adequate protection against damage on site is to be provided by the

Figure 11.7: House under construction, Bungendore, New South Wales. Note the large stone lintel and light switch chased into the wall.

general contractor. All stones are to be checked for damage on arrival at the site and stored in sequence for fixing.

12. Protection and cleaning down: On completion of the work, the masonry contractor shall thoroughly clean down the work and attend to any defects in the pointing, before handing it over to the general contractor who will then assume responsibility for the condition of the work.

If there is no architect, the person responsible for ensuring quality control can amend the specifications accordingly. Some decisions will need to be made by the proprietor (usually the owner), and this should be written in if there is no other party acting for them. Specification notes in general, and for the other trades in the construction works, are also required, but are not given here.

Building stonework can be time-consuming. If all goes well, a stone house can be constructed in a timeline of a year or two. However, if other commitments are more pressing for the owner builder, life goes on, and the stone walls may wait. Luckily, stone walls are one part of a building site that can remain exposed to the weather without much deterioration. Other materials are

Figure 11.8: Tall chimneys with 'cricket' gables, Hobart, Tasmania.

more vulnerable to the forces of nature, such as wind, water, sunlight, fungus or insect attack.

Large stone buildings require time and perseverance. Differences between stonework finishes and decorative details need to be carefully documented and accurately resolved on site (Figures 11.6 to 11.9). The great European cathedrals

Figure 11.9: Rock-faced arch and corner with walling smoothed, corbelled chimney, Hobart.

were built by itinerant teams of master stonemasons, who downed tools when the money ran out. When more money was available, another team was employed. On many sites, the fashion of ornament had changed, so the two styles of stonework sit side by side where the work carried on. As for fortified castles, they were being built to an unknown deadline, the date of the next enemy attack. Sometimes they missed the deadline, such as at Chateau Gaillard in France, where the attackers found a drain and entered the inner keep, after storming the unfinished outer fortifications of the castle (Deville 1829).

Contract documentation is the means to build exactly what is drawn, within a set time and to an agreed price. The process of putting drawings out to tender and administering the building contract is the work of the architect. Any changes to a contract come at a cost, because the contract is based on the agreed drawings and specifications.

The contract can also be between the client and the builder, but this form of contract has no assurance of contract administration to ensure the work is done exactly according to plan. In this case, the onus is on the client to ensure the building work is done to their satisfaction, through a good relationship with the builder.

Appendix 1: Stone footings

With appropriate conditions, stone or brick footings may be used. The foundations must be suitable, offering stability, such as solid bedrock. On a foundation of rock or stable soil, stone footings can be built using the old methods. If the stones being turned up on the site are too hefty to use higher in the wall, they could be useful for footings. The presence of historic buildings in the vicinity can give an indication of the suitability of the local stone. Today's regulations (*Building Code of Australia*, Housing Provisions 2011) allow for alternative construction methods, on a performance basis, as acceptable construction practice. The information in the regulations is only for concrete footings, but the use of other materials is at the discretion of the building surveyor. So for sites with stable soil, if stone footings are wanted, it is worth discussing them, as the council may accept them. If there is uncertainty about the footing design, it should be referred to a structural engineer.

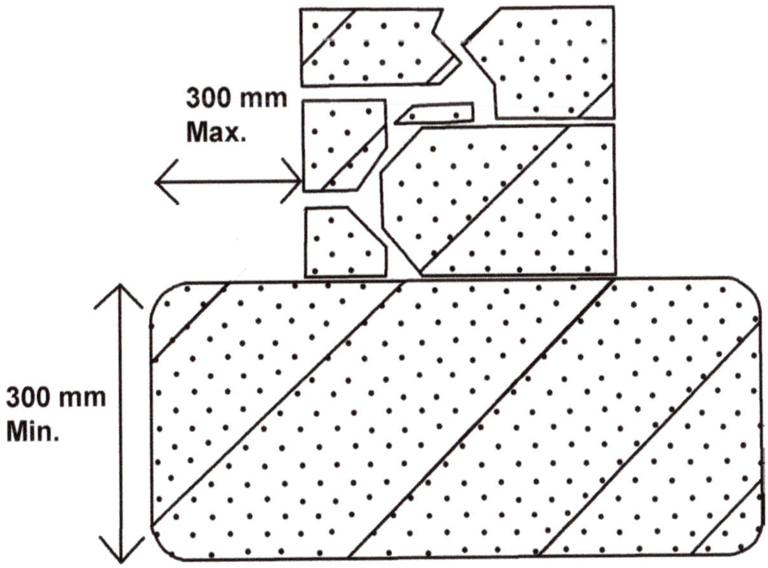

Figure A1.1: Stone footings, rule of thumb.

A rule of thumb for stone strip footings, as an alternative to reinforced concrete, can be found in the outdated Victorian building regulations (*Uniform Building Regulations, Victoria* 1969) or in footing proportions from the ordinance for New South Wales (Watson Sharp 1946), which allowed them be used on non-cohesive soils or rock foundations for buildings up to two storeys high.

The minimum thickness of the base course should be 300 mm (12 in.) and the stone footing should not project more than 300 mm (12 in.) from the face of the wall or pier (Figure A1.1). The stones should be large enough to provide for only one stone across the footing width, and laid in their natural beds in cement or composition mortar.

Other recommendations suggest that the width of the footing should be twice the width (or thickness) of the wall. The actual width needed for the footing to distribute the load depends on the bearing capacity of the foundation soil. As with concrete footings, the top of the stone footings should be level. Stepped footings may be required to accommodate sloping ground.

Appendix 2: Retaining walls

It is not unusual for sites providing abundant rock for building to also be on a slope. The creation of level outdoor areas therefore requires the construction of retaining walls. If the wall has a low profile, it can create the illusion, from above, of continuous grassland, with little indication of why the sheep do not venture higher up the hill. A retaining wall of this type is known by the Scots as a ha-ha.

The weight of a thick stone wall can be enough to retain the earth from uphill with just the force of gravity, but only to a given height. This height can be determined if enough is known about the materials and loading. If the height of earth to be retained is greater, engineered footings and reinforcement are required, to ensure that the wall does not tip over. Stone retaining walls must be able to

Figure A2.1: Curved retaining wall of cellar under construction, house near Hepburn Springs, Victoria.

Figure A2.2: Dry stone being built into a retaining wall on a steep hillside, Castlemaine, Victoria.

withstand lateral pressure from uphill, caused by the soil and the water contained (Figure A2.1). To reduce the water pressure, a mortared stone retaining wall needs to drain water out from the lower level through weep holes or drain pipes. Dry stone retaining walls can drain out water between the stones.

A dry stone retaining wall should have a slight angle of batter against the hillside. If the wall is to retain new soil fill, it can be built course by course as the soil is dumped behind it (Figure A2.2). The stones should be carefully packed together and tamped down before the next course and the next layer of soil are added. If it is available, shale is a useful stone to pack in behind the back of the wall to level it for the next course.

Retaining walls built with stone and mortar, and/or concrete are of three main types: (1) gravity, (2) buttress (or counterfort) and (3) cantilever (Figure A2.3). These walls should act as a single unit, using mortar and reinforcing as required. A backfill of rubble over a drain at the base is necessary to reduce water pressure on the wall, and helps to minimise pressure from soil movement, especially if the soil is expansive clay. These methods can also be applied to brick or reinforced concrete blockwork.

In stonework, gravity retaining walls are the most common. These depend entirely on the weight of the masonry for stability.

Buttress or counterfort retaining walls are restrained by vertical buttresses on the outside, or counterforts on the inside, providing lateral support so that the wall

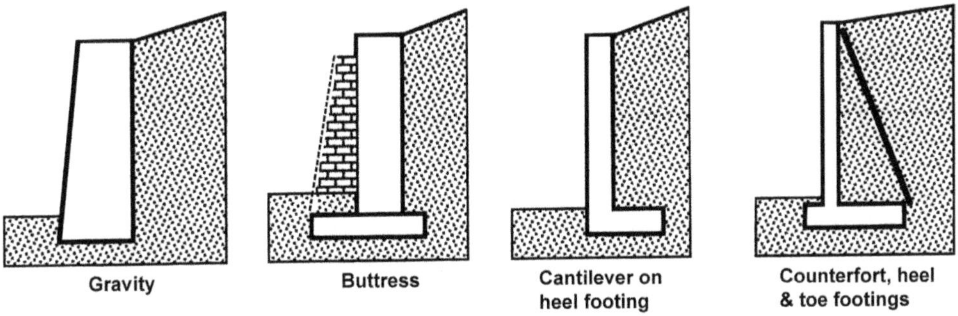

Figure A2.3: Retaining wall options.

spans between them. The wall may need to be built with reinforcing bars to take the pressure from the soil. These are used for high retaining walls.

A footing much wider than the wall can make its base rigid enough for the wall to act as a cantilever from the bottom. It is restrained from bending either by a heel of concrete under the retained soil, or a toe at ground level on the low side, or both. Continuous reinforcing bars are placed in the footing and up through the wall.

The size of a gravity retaining wall can be worked out by geometry if the characteristics of the retained soil are known (see Figure A2.4). The sideways thrust acting on the retaining wall is a result of the sliding tendency of the soil behind it. If the wall were not there, the top part of the fill would slip down. The pressure on the wall from this is caused by the weight (*w*) of a triangular prism of earth above the plane of rupture. This plane of rupture is taken to be midway between the angle of repose (the natural incline for the soil type) and the vertical plane of the wall.

The theoretical direction of reaction against the force *w*, from below, is perpendicular to the plane of rupture. However, friction affects this direction and the reaction force line is brought upwards by an incremental angle equal to the

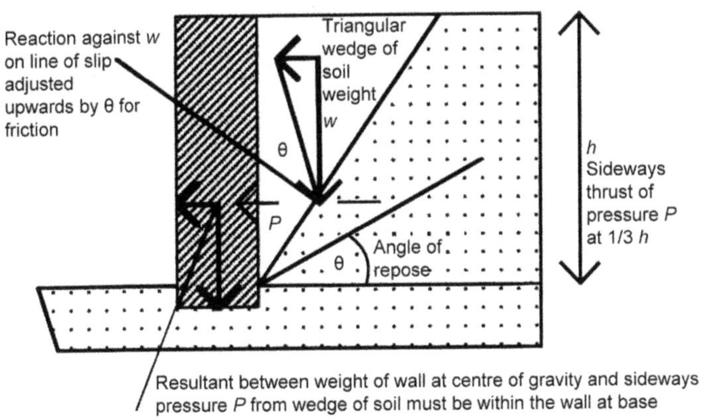

Figure A2.4: Soil loads.

angle of repose. The resultant force between *w*, the force of gravity on the soil weight, and the reaction at the adjusted angle is taken to be *P*, the horizontal pressure acting to tip the wall sideways. It occurs at a height of 1/3 *h*, where *h* is the height of the retaining wall (or higher if it retains soil at a slope).

A graphical method can be used to determine whether the wall thickness is adequate. The wall should first be drawn to scale in cross-section. A line is then drawn vertically through the centre of gravity of the wall, and the weight of the wall is scaled on this line. The differential weight by force of gravity on the material in the wall and in the soil are roughly proportional to their comparative cross-sectional area. The force *w*, and the force pushing outward from the friction along the plane of rupture, is scaled to show a dimension for the horizontal force *P*. The angle of repose is used to find the resultant angle between these two forces. The pressure *P* acting on the wall is drawn to the same scale and the resultant diagonal must fall within the thickness of the wall at the base.

If the soil slopes up from the top of the retaining wall, then the centre of gravity moves up too, and the position of the resultant horizontal force *P* is located higher.

In this case the height of soil retained, for sloping soil, *h*, increases to the upper edge of the triangular prism formed with the plane of rupture.

The value of the angle of repose will vary but can be taken as:

- 25° for loose sand
- 30° for sandy soil
- 35° for granular soils
- 45° for drained clay, and
- 48° for gravel or rock fill.

A form of gravity retaining wall that can be built of dry stonework is the crib wall. Built in much the same way as a precast concrete crib wall, the face of the wall has a slope of up to 1 in 8. For low walls using through stones, the face can be almost vertical. The fill behind the crib wall should be of broken or uneven rock, to allow water movement, and should extend 900 mm behind the wall stones. Each course should be securely fitted in place and stable, with the fill packed firmly behind, before the next course is added.

Appendix 3: Arches

The evolution of the arch

The arch is a wonderful thing, for its ability to span across wide openings with a row of stones, where the work of getting a single stone up there would be back breaking (or require the expense of a crane). Unlike a lintel, the arch does not just sit on top of its supports. The forces between the individual stones in an arch must be counteracted at the base and sides. Mortar between the stones will help to bind them together, but the arch still relies on the right amount of masonry in the right places to avoid cracks forming.

From personal experience, when we took a large stone lintel across the width of our cellar window it was hard work, with a lot of setting up. As it was close to the ground, we were able to get the lintel into place by leverage (see Figures 4.14 and 4.15). We manoeuvred the stone off the back of a ute onto a temporary ledge supported by a couple of 44-gallon drums. Once the stone was resting above the wall, the mortar bed was put in place for the bearing ends, then the planks holding the lintel stone were removed, so it was resting on the mortar. The whole process took so much effort! We were elated when we achieved it successfully.

After this, we did not want to use lintels for the openings higher up. We could not easily imagine getting large stones up there. However, we were concerned about whether we could get enough stones suitable for shaping to do arches over all the windows. During the process of pondering this, when Doug and I were going to the beach one day, we saw a stone arch built of rough rubble stone in a public park in Mornington (Figure A3.1). This gave us hope that we could make arches with found stones, without the need to shape them very much at all. The search was on then, for voussoir-shaped stones. We built arches from naturally shaped voussoirs for the two main levels of our stone tower.

Over the course of history, arch styles became refined, often combined with the classical orders of architecture dating back to Greek temples (for which the Greeks

Figure A3.1: Just after the break-through moment when we saw that it is possible to build arches of rubble stone, here was the proof that gave us inspiration, Mornington, Victoria.

used lintels). The ancient Greeks were familiar with the arch, and used it in engineering. In civic architecture, or temples, where they sought perfection, columns supporting lintels were preferred. Perhaps this was for the benefit of having corners supported only by columns, to bring in more light, since the arch requires a width of wall to buttress it against a tendency to push outwards on either side, which the lintel does not.

The columns used in the classical orders of architecture are round, which reduces the area of light blocked by construction. However, they each need a square capital to spread the load to a square wall again for bearing the ends of a lintel. The clear span of a lintel is limited by the length of the available stone. An arch can be used to increase the width of the opening. The combination of arch and column requires balance, as a thin column is not sufficient to counteract the forces of an arch.

The classical Renaissance architect Palladio wrote in his *Quattro Libri* (four books on architecture) 'Doric will always be placed under the Ionic, the Ionic under the Corinthian'. After our own experience with a stone lintel, I would add that heavy stone lintels, which suit the Doric style, can be used at the lower level, but to contemplate them higher up may only add back-breaking expense to the job. As the construction goes higher, the proportions of structural supports can become more slender. Lintels have an advantage over arches, in that they do not cause the spreading effect that puts a lateral thrust onto the arch supports. Arches must always be used with sufficient wall width on either side to counteract this thrust. An arcade is a series of arches in which the lateral thrust of each arch counteracts the next. At the end of an arcade, this force must be resolved.

In a number of his buildings, Palladio used a technique that combined a lintel on either side with a semicircular arch. This treatment combines the advantage of using some larger stones for lintels and reducing the arch width and height to manageable proportions. At the same time, the lintels carry the arch thrust to the solid wall at the sides of the full opening width. With this combination, any variations in the width of the opening can be accommodated in the lintel span, so the arch size can be kept consistent.

Arch geometry

The semicircular arch and the segmental arch are both based on the geometry of a circle. While the semicircular arch is appropriate for narrower openings (Figure A3.2), the segmental arch is a logical way to widen the opening without increasing the rise of the arch (Figure 8.20).

A camber arch has a curve under the arch, but it has a minimal rise (Figures 6.5 and 9.3), and without much of the advantage of height difference to transfer the loading, it is not much more than a flat arch. For a flat arch or a camber arch, if the

Figure A3.2: Bold voussoirs over archway, Cassilis, New South Wales.

arch stones will not form enough of a wedge shape to lodge them firmly in place, a steel arch bar should be used.

An oval or elliptical arch can be set out using string, a compass or specialised tools, each with different results (Figure 1.13). Pointed arches are derived with an arc on each side generated from a different centre point (Figure 7.22). The curves of catenary or parabolic arches are a departure from the geometry of a circle, and can be set out by geometry or by other methods to replicate the forces of gravity.

Semicircular arch

The semicircular arch has a simple geometry, where the midpoint of the span determines the centre of curvature. The radius, which is the rise of the arch, is half the width of the opening, and the joints between stones radiate from the centre point of the springing line. The curve is divided into parts, as equal as possible, of uneven number, so that there is a keystone located centrally at the top to balance the loading.

Segmental arch

A segmental arch has a rise of more than 1/12 the width of the opening.

When the height of an arch is less than the radius of a circle of the same curvature, the arch stones can align either to the centre of radius, or to the point at the centre of the springing line of the arch (Figure A3.4). To establish the centre of curvature of an arch of a given span, the diameter can be worked out for the curve from its height. This may be referred to as the developed diameter, and it is given by the formula: $D = (L^2/4h + h)$, derived as follows.

Developed diameter

To find D, the developed diameter for a segmental arch, where L is the length of the arch span and h is the height of the arch.

The formula for a circle with its centre at the origin is: $x^2 + y^2 = r^2$, where r is the radius of the circle. For a point on the circle (x, y), x is half the length of the span L, and y is less than the radius r by an amount equal to the height h (Figure A3.3).

$$\text{so } \left(\frac{L}{2}\right)^2 + (r-h)^2 = r^2$$

$$\text{then } L^2 = 4\left(r^2 - \left(r^2 - 2rh + h^2\right)\right) = 4\left(2rh - h^2\right)$$

$$\text{so } 2rh = \frac{L^2}{4} + h^2$$

now $2r$ is the diameter, D, therefore:

$$D = \frac{L^2}{4h} + h$$

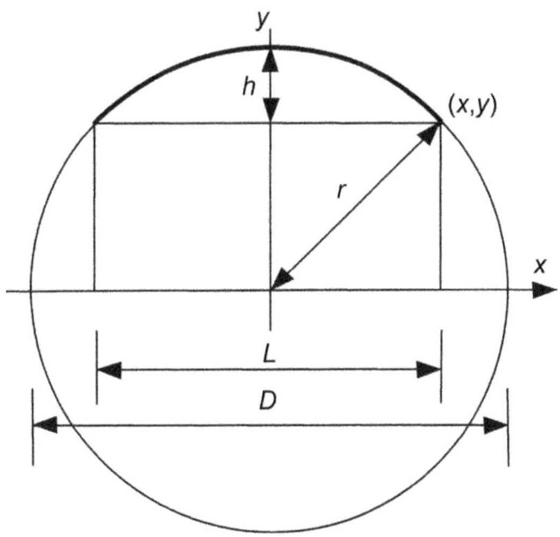

Figure A3.3: Derivation of the developed diameter for a segmental arch.

For arches of minimal rise supported by steel, the camber or curve of the arch bar can be determined using the formula for the developed diameter. The arch bar should be supported at either side of the opening with a bearing width of at least 100 mm. Tables for brickwork openings using arch bars or lintels are useful to determine the size and bearing of arch bars.

Oval arch

One way of making an oval arch is from the radii of three circles. The arcs of two small circles are set out, with a segmental curve across them, meeting each side arc

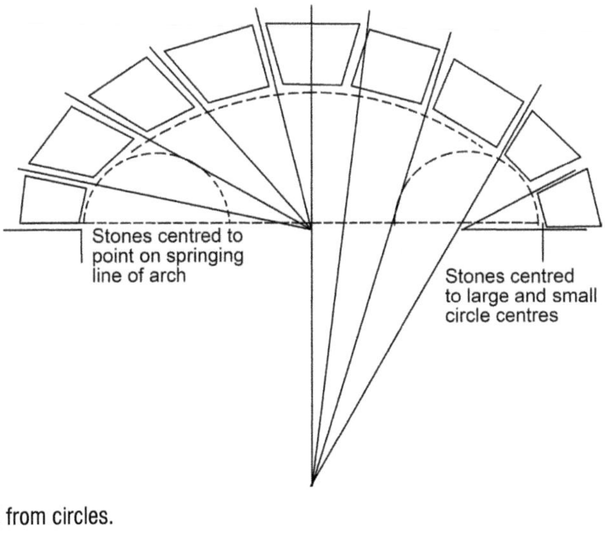

Figure A3.4: Oval from circles.

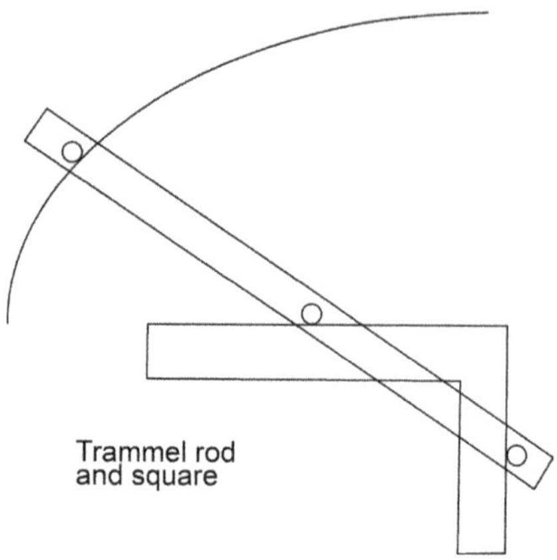

Figure A3.5: Trammel.

at a common tangent. The voussoirs can be shaped with their sides radiating from the mid point of the springing height. Or alternatively, the centre of radius for the voussoirs can be adjusted between three points, determined by the location of the common tangent of the larger and smaller arcs.

Elliptical arch

The true ellipse shape is determined with focal points, not circles. The shape of an elliptical arch can be laid out using either a string or a trammel rod and square, giving a different result from the oval arch above.

The trammel rod is a flat stick with a pencil hole at one end, for marking the curve, and at the other end it has two pins located one for the rise, and one for half the span. The pins slide along the edges of a square, so the ellipse curve can be marked with a pencil (Figure A3.5).

The string line is a simpler method (Figure A3.6). It can be laid out roughly using a string extended from two points along the springing line, equidistant from the centre line. Adding the distances from any point on the inside of the arch to the

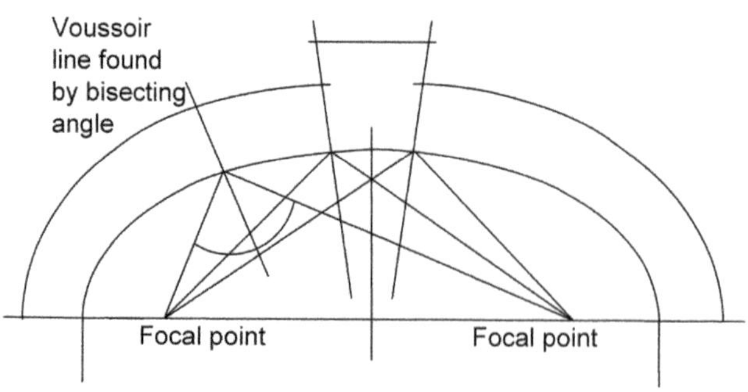

Figure A3.6: String method.

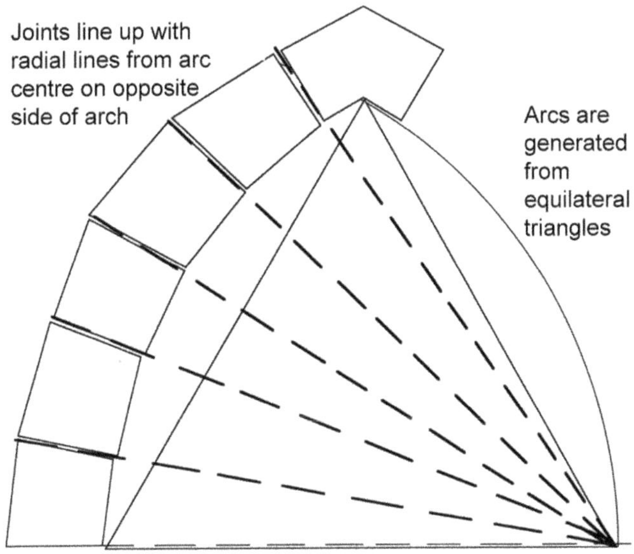

Figure A3.7: Gothic arch.

two focal points gives the same result (the length of the string). This curve can be then divided into segments to establish the locations of the voussoirs. To determine the shape of the stones, the joint lines are taken from the mid-line of the angle formed by the string at each segmental point.

Pointed arches

A pointed or gothic arch can be set out from an equilateral triangle. The curves are taken as an arc on either side, meeting with a point at the top. In this geometry the keystone is concave, with an indented point at its base, unless it is fitted as a wedge coming down to a point at the top of the arch (Figure A3.7).

The arch geometry can be varied by using a taller or shorter triangle for a taller arch, or a squat arch. The location of the centre of the radius for the voussoir joints is moved to suit the curve. For the taller, or lancet arch, the radial lines are generated on the far side of the arch opening. A more squat arch, or drop arch, has its radius shorter than the arch span (Figure A3.8). The Tudor arch follows a shape generated by circles half the arch width, meeting an arc for the centre portion, which stems from a common tangent with the smaller circle on each side.

Catenary or parabolic arch

The catenary arch has a more rounded curve at the top than at the sides. The shape of the arch can be set up to follow the shape of a suspended rope. The rope responds to gravity and its own self load to form a natural curve in tension. This curve turned upside down lends itself naturally to the loads of compression.

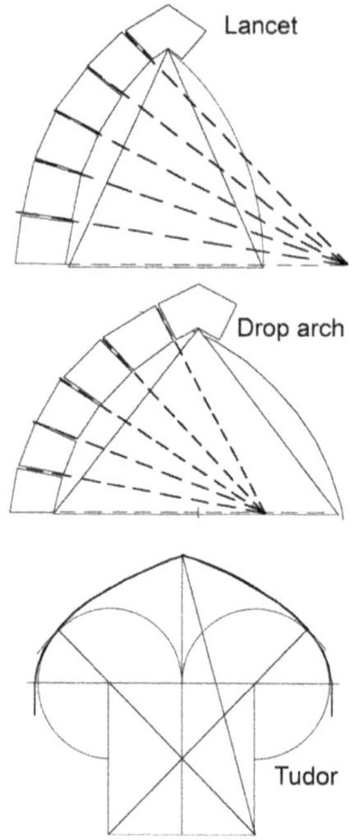

Figure A3.8: Pointed arch variations.

A parabolic curve can be worked out using the mathematical formula for a parabola. The proportions are such that for each unit out from the centre, the curve is heightened by a progressive increment (Figure A3.9). In the case of the basic curve $y = x^2$ the height gets larger by a regular succession of odd numbers. To change the proportions for better fit, the equation can be adjusted by the appropriate factor.

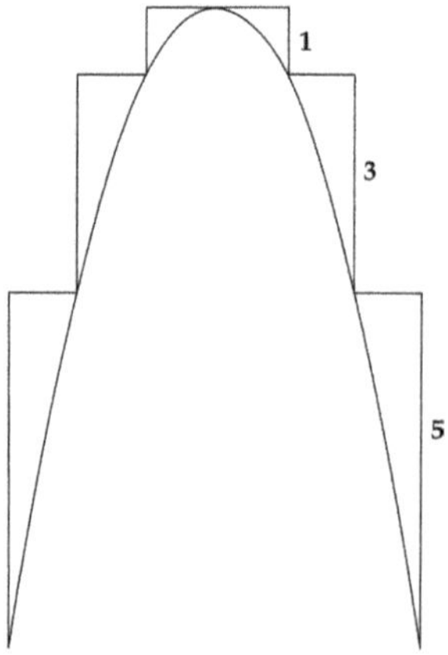

Figure A3.9: Parabolic curve.

Glossary of masonry terms

abutment: solid supporting wall at end of arch or vault, such as for a bridge

arch: curved masonry over an opening, made of bricks or shaped stones (voussoirs)

arris: edge where two flat surfaces of stone meet

ashlar: masonry using precision worked stones with close joints

arcade: series of arches supported by columns or piers

aggregate: hard fragmented material (rock pieces or sand) for mixing into mortar or concrete

battered: (of walls) leaning in, often as retaining wall

bed: (1 – quarry) sedimentary plane of rock stratification; (2 – building) mortar layer prepared for laying a stone

bedding plane: parallel layer in sedimentary rock where sediment originally settled

bevel: splayed angle on a worked stone; a tool for marking out angles

block stone: large stone block roughly squared at quarry

bluestone: basalt of blue-grey colour (usual term in Victoria for local basalt)

boaster: chisel with wide cutting edge, up to 75 mm; boasted: tooled finish made with boaster

bonder: stone laid to increase strength of stonework, usually across width of wall

broach: chisel used to work narrow surfaces

brought to course: laying tops of stones to level, with smaller stones stacked to match level

buttress: masonry pier bonded to a wall, often thicker at base, built to counteract lateral thrust

capital: stone at the top of a column or pilaster, decorative

cavity wall: double wall construction, two thicknesses of masonry separated by air gap

centering: method of framework to give curved shape, usually made of timber, upon which arch, dome or vault is built

chamfer: flat splayed edge between two flat surfaces

cladding: non-load-bearing wall facing, thin slabs of stone (or other material) used vertically

composition mortar: cement-lime mortar using (in parts by volume) from 1:3:12 to 3:1:12, but usually 1:1:6 (cement:lime:sand)

coping: (noun): top course of a wall, designed for weather protection, may be built from more impervious stone; (verb): technique of breaking stone, grooved on each side, worked with chisel along groove

corbel: projecting masonry stepped out in courses with at least three quarters of each projecting stone bedded on course below, used for structural support

cornice: course of stone projecting from wall; cast plaster fixed to wall under ceiling

course: horizontal layer of masonry units along length and width of wall

damp proof course (DPC): physical barrier in mortar to prevent moisture flow into wall from groundwater or rainwater

dimension stone: a block of quarried stone for cutting to specified size, the stone when cut

drip mould: stone profile with small projection under lower front edge, usually grooved behind, projecting beyond other wall stones, for water to drip from

efflorescence: encrustation of crystals on wall, from exposure of dissolved salts drying at surface

extrados: upper or outer line of stones formed as an arch, on which the upper wall is built

face: the side of a stone that is visible in a finished wall

face bedded: (e.g. for sedimentary stone) laid perpendicular to its natural bedding plane (can cause weathering problem)

flat arch: shaped stones or bricks splayed to spread load to sides, with horizontal bed at top for next course (extrados), usually with slight camber to underside (intrados)

footing: lowest part of construction, which transfers the structural loads to the foundation

foundation: the ground supporting the building

freestone: stone that will split equally well in any direction

grout: weak mortar, for purpose of filling up and finishing off gaps between stones in a wall

header: stone or brick laid with short face showing, and often bonding wall front to back

hearth: floor of a fireplace, extending beyond fireplace opening with fireproof construction

initial set: early stage of mortar hardening, before developing strength, when it can be cleaned

intrados: interior soffit of arch or vault

joggle: a connection piece built into a recess in a stone to stabilise against lateral movement

joint (in rock): natural cracks or lines of weakness where rock is likely to break; (in stonework): mortar infill where stones are placed next to each other; (bed joint): line of mortar in the wall between successive courses of stonework; (decorative): various ways of finishing exposed face of mortar between stones (e.g. flush joint, raked joint, struck joint, tuck pointing)

keystone: top stone of an arch, acting as a wedge to transfer downward thrust to sides

lap: overhang of a stone where it sits beyond the stone directly below (e.g. at a break in work)

lava: surface-cooled volcanic rock, flow of molten rock or ejaculate from volcano

lime: a calcium oxide used in composition mortar or lime mortar, which reacts with water to form slaked lime (calcium hydroxide)

lintel: beam (stone or otherwise) across a door or window opening

magma: molten rock below the Earth's crust, from which igneous rocks are formed

masonry: stone, brick, concrete block or similar construction, laid unit by unit, and set in mortar

movement joint: (also expansion joint, control joint, articulation joint) straight joint filled with compressive material to allow movement in the wall, required in brick walls over 6 metres long, to allow for the expansion that occurs over time; contraction of concrete necessitates joint lines (cracks will form otherwise); note: stone is generally more dimensionally stable

natural bed: surface parallel to the layers in a stone resulting from its original sedimentation

niche: small recess in a wall, usually arched

oriel window: small projecting window, usually three-sided, supported on corbel in upper wall

parapet: a portion of wall carried up above the junction of the wall with the roof (or gutter)

perpend: vertical mortar joint on wall face, end face of stone or brick (generally when built through double brick wall)

pier: vertical masonry element of structure, usually square in cross-section or attached to wall

pilaster: pier with minimal projection from wall, structural and sometimes decorative

plug and feathers: specialised tool, made up of chisel and wedges, for splitting stone

plumb: vertical alignment as indicated by a weight (plumb bob) on a string (plumb line)

pointing: filling gaps in mortar joints, smoothing off defects with tool or brush

quarry sap: natural moisture containing minerals, which harden and crystallise on drying

quoin: external corner stones, especially where larger than other wall stones or finished differently

relieving arch: curved masonry in a wall above a flat opening to reduce loading on the lintel

retaining wall: masonry built for the purpose of preventing lateral movement of earth or fill

rubble: stone of irregular size and shape

rusticated: built in large textured stone blocks with deep joints or margins, usually at base of wall

scoria: cellular lava rock of high density, formed from the froth of volcanic action

spall: fragmentation of thin pieces or stone chips peeling away from a stone surface

straight joint: two or more perpends, each directly above the other, in successive courses (this is to be avoided unless allowing for a future opening or at a movement joint)

soffit: underside of a building component

springer: stone at base of arch where curve rests on wall

stretcher: stone or brick laid with its length parallel with the direction of the wall

stone dust: fine sand of same type as stone, for thin mortar joints with best matching quality

tailings: leftover stone from a previous quarry or mining activity

through stone: a stone going the full thickness of the wall (see also perpend)

toothing: projection of alternate or occasional stones at end of wall to allow bonding of future work into existing courses

trap: any of various fine-grained dark igneous rock especially basalt, often used for road-making

tympanum: curved area under arch and above lintel or spring line along base of arch

underpinning: new footings under existing wall for adjacent excavation to reduced levels

veneer: non-structural wall surface or skin of masonry

voussoir: shaped stone block used to build the curve of an arch

wall ties: connecting strips, usually wire, embedded in mortar of cavity wall

weep holes: (at base of cavity wall) perpends left open to let moisture drain out

wrought: worked or shaped by chisel to correct dimensions and face profile (of stone)

Endnotes

1 For example, cladding failure in the USA, Gail Hook, 'Look Out Below', pp. 58–61, *Progressive Architecture*, Feb 1994. The article describes the expensive replacement of facing stones on the 82-storey Amoco Building, built in Chicago in 1973. The original Carrara marble, from Italy, bowed due to greater expansion on its external face. It was replaced by slightly thicker granite from North Carolina, found to be stronger initially and over time than the marble. Problems also occurred on the MMBW building in Spencer Street, Melbourne in the 1970s, originally faced with basalt from Warrnambool, replaced with anodised aluminium. The cracking of the stone may have partly been due to the combination of blasting in the quarry and the cutting of it into a thin veneer, leaving unresolved weaknesses.

2 Information on the Japanese cutting technique of using a curved V-incision is from personal communication with Graham James, who drew a quick sketch for me.

3 Colour issues were noticed in Pyrmont stone stockpiled for restoration work when the 'yellowblock' quarry site there was closed for urban development. The colour change due to weathering was not occurring when the large stockpiled blocks were cut for building stone. The chemicals in solution that cause the colour change were less active, and the stone did not consistently match the desired heritage stone colour as anticipated (Swann 2008; Powell 2008).

4 In *Small Houses, Their Economic Design and Construction* (Flagg 1922), Flagg's module length is 3 ft 9 in. (1140 mm) on the basis that standard timber lengths are in 2 ft (610 mm) increments. Standard lengths are used (e.g. 2.4 m, 3.6 m, 4.8 m) with a deduction of 150 mm for bearing on walls each end.

5 Author's note: we were able to get two arched windows from a demolition site. We took them from a double brick wall with an arch on the outside and a timber lintel with relieving arch on the inside. They appear in Figure 8.17 on the left with a temporary roof over them (no arches yet).

6 In Loraine Kuck's *The World of the Japanese Garden, From Chinese Origins to Modern Landscape Art* (Kuck 1968), on p. 93 she refers to a historic document, *Sakuteiki*, which emphasises the importance of rock placement in early Japanese gardens.

References

Books and journals

Aitken R, Jones D and Morris C (2006) *Adelaide Botanic Gardens Conservation Study*. Department of Environment and Heritage, South Australia.

Anderson B (1977) *Solar Energy: Fundamentals in Building Design*. McGraw-Hill, New York.

Ashurst J and Dimes FG (1977) *Stone in Building: Its Use and Potential Today*. Architectural Press, London.

Australian Building Codes Board (2011) *Building Code of Australia (BCA), Volume 1 Housing Provisions (Class 1 & Class 10 Buildings)*. Australian Building Codes Board, Canberra.

Baglin D and Austin Y (1976) *Sandstone Sydney*. Rigby, Adelaide.

Baker RT (1915) *Building and Ornamental Stones of Australia*. Technical Education Series No. 19. WA Gullick Government Printer, Sydney.

Barran FR (1962) *Der Offene Kamin: The Open Fireplace*. Hoffman, Stuttgart.

Deville A (1829) *Histoire de Chateau Gaillard, et du Siege qu'il soutint contre Philippe-Auguste en 1203–4*. Edouard Frère, Rouen.

Flagg E (1922) *Small Houses, Their Economic Design and Construction*. Scribners, New York.

Fleming J, Honour H and Pevsner N (1972) *A Dictionary of Architecture*. 2nd edn. Penguin, UK.

Foster JS (1973) *Structure and Fabric Part 1*. Batsford, London.

Geological Survey, Western Australian Department of Mines (1984) *A Guide to the Building and Facing Stones of Perth and Fremantle*. Western Australia, Department of Mines, Perth.

Giovannini J (1983) Architecture: Charles Foreman Johnson. Arizona. *Architectural Digest*, February.

Gunn Swing S (1982) A stone tower. *The Owner Builder* **4**, 42–45, June.

Hook G (1994) Look out below. *Progressive Architecture* **75**(2), 58–61, Feb.
Howe MA (1915) *Masonry.* Wiley & Sons, New York.
Johnson DL (1977) *The Architecture of Walter Burley Griffin.* Macmillan, Melbourne.
Kern K (1974) Heating and cooking with wood. In: *Producing Your Own Power: How to Make Nature's Energy Sources Work for You.* (Ed. CH Stoner) pp. 103–135, Rodale Press, Emmaus, PA, USA.
Kern K, Magers S and Penfield L (1976) *Owner Builders Guide: Stone Masonry.* Owner Builder Publications, California.
Kern K and Magers S (1978) *Owner Builders Guide: Fireplaces.* Owner Builder Publications, New York.
Kincaid P (1991) *Traditional and Modern Building with Stone.* Hale & Iremonger, Sydney.
Kuck L (1968) *The World of the Japanese Garden, From Chinese Origins to Modern Landscape Art.* Weatherhill, New York.
Langmead D (2003) *Frank Lloyd Wright: A Bio-Bibliography.* Praeger, USA.
Lloyd C (1954) *Building Construction for Craftsmen and Builders.* Macmillan, Melbourne.
MacDonald G (1872) *The Princess and the Goblin.* Dent & Sons, London.
Maxlow J and Maxlow A (1984) *Rock Solid: Building a Stone House in Australia.* Night Owl, Victoria.
Mines Department (1949) *Victorian Building Stones.* Mines Department, Melbourne.
Mitchell CF and George A (1904) *Brickwork & Masonry – A Practical Text Book for Students, and Those Engaged in the Design and Execution of Structures in Brick and Stone.* Batsford, London.
Molloy E (1946) *Brickwork and Masonry.* 3rd edn. George Newnes, London.
Nearing H and Nearing S (1970) *Living the Good Life: How to Live Sanely and Simply in a Troubled World.* Schocken Books, New York.
Nickey JM (1979) *The Stoneworker's Bible.* TAB Books, USA.
O'Neill H (1965) *Stone for Building.* Heinemann, London.
Peters FF (1936) *Houses of Stone.* Putnam, New York.
Peters FF (1949) *Pour Yourself a House.* Whittlesey House, New York.
Pfeiffer BB and Nordland G (Eds) (1988) *Frank Lloyd Wright: In the Realm of Ideas.* Frank Lloyd Wright Foundation, USA.
Powell R (2008) Yellowblock, to stockpile or not? *Discovering Stone* **13**, 30–31, March.
Purchase WR (1896) *Practical Masonry: A Guide to the Art of Stone Cutting.* Crosby Lockwood & Son, London.
Rainsford-Hannay F (1957) *Dry Stone Walling.* Faber & Faber, London.

Roberts R (1964) *Your Engineered House.* M. Evans and Company, New York.

Ruskin J (1896) *The Ethics of the Dust: Ten Lectures to Little Housewives on the Elements of Crystallization.* 9th edn. George Allen, London.

Schwenke K and Schwenke S (1975) *Build Your Own Stone House, Using the Easy Slipform Method.* Garden Way, Vermont, USA.

Stowe HB (1868) *Household Papers and Stories.* Houghton Mifflin and Company, Cambridge, MA, USA.

Swann J (2008) Sydney Yellowblock, is its revival a blessing or a curse? *Discovering Stone* **13**, 22–27, March.

Swing S (1984) Stonework in housing. In: *Owner Building and Earth Architecture, Proceedings of National Conference 3–5 February, University of Melbourne.* (Ed. T Howard) Dept. of Architecture & Building, University of Melbourne and Ministry of Housing, Melbourne.

Taylor R (1990) Revealing masons' mysteries. *Architects Journal* **26**, Sept.

Turnbull J and Navaretti PY (1998) *The Griffins in Australia and India: The Complete Works and Projects of Walter Burley Griffin and Marion Mahony Griffin.* Miegunyah Press (Melbourne University Press), Melbourne.

Tuulse A (1958) *Castles of the Western World.* Thames & Hudson, London.

Twidale CR (2000) Granite outcrops, their utilisation and conservation. *Journal of the Royal Society of Western Australia* **83**(3), Sept.

Uniform Building Regulations, Victoria (1969) Government Printer, Melbourne.

Vaux C (1857) *Villas and Cottages.* Harper & Brothers, New York.

Warland EG (1953) *Modern Practical Masonry.* 2nd edn. Pitman, London.

Watson Sharp W (1969) *Australian Methods of Building Construction.* Angus & Robertson, Sydney.

Watts WW (1899) *Geology for Beginners.* Macmillan, London.

West D (1990) Selection of stone and quality control. In: *Proceedings, a New Stone Age in Architecture, Dimension Stone in NSW.* 29 November, Sydney. Department of Minerals and Energy, Sydney.

Whitten DGA with Brooks JRV (1972) *The Penguin Dictionary of Geology.* Penguin, UK.

Willingham A (1987) *Narrapumelap and the Lions of Wickliffe.* Historic Buildings Council, Victoria.

Wright FL (1928) In the cause of architecture, III: the meaning of materials – stone. *Architectural Record*, **63**(4), 350–356, April.

Wyss JD (1821) *Swiss Family Robinson.* 4th edn. MJ Godwin & Co., London.

Websites

<www.artsoutwest.org.au> – organisations – Abercrombie House

<www.castlecrag.org.au/builtenvironment/builtEnvironment.htm> – Castlecrag Community, Heritage

<www.environment.sa.gov.au> – Botanic Gardens – *Adelaide Botanic Gardens Conservation Study*

<gen.lib.rus.ec> – *Frank Lloyd Wright: A Bio-bibliography*

<www.gutenberg.org/files/31217/31217-8.txt> – Harriet Beecher Stowe (1868)

<www.nattrust.com.au> – database – Narrapumelap

<adbonline.anu.edu.au>

<www.percocomarble.com> – natural stone selection

<www.pir.sa.gov.au/minerals/geology/mineral_resources/commodities/dimension_stone>

<www.rswa.org.au/content/work/journals/PDF/83(3)/v83(3)twidale.pdf> – Twidale, C.R. (2000)

<www.smh.com.au/travel/travel-factsheet/bathurst-20081113-5yl9.html> – Abercrombie House and Strath

<www.soapstonesculpture.com/soapstone.html> – Sandy Cline, 'Properties of Soapstone'

About the photographs

The photographs were all taken in Australia.

Most of the photos are by the author. Exceptions include:

Figure 2.6, photo by Doug Swing (and perhaps a few others from our collection).

Figures 3.11, 6.24, 6.25, photos by Jane Millar.

Figures 6.2, 6.18, photos by Ted Howard.

Figure 8.20, photo by Nat Beloborodova.

People in the photographs

Figure 7.24 – my mother, Pat Gunn (née Gibson) was an avid collector of building stone. With collected stone, she had a high front wall and two garden walls built at the Melbourne home of my childhood. She collected marble and granite offcuts from monumental masons, to use as paving. This was just the start. She went on to have two stone houses built from nothing but a rocky piece of land, first near Hepburn Springs, Victoria, then near Moruya in New South Wales. She asked me to help with the design of the first, when I was still a student. It has taken longer than I could have anticipated, and I am still learning.

Figures 4.4, 4.8, 4.14, 7.7 – Harry Vanderburg met Pat when he was a swimming pool tiler. Their friendship led to him paving around her new pool with marble offcuts she had collected. Later he went with Pat to work on the house near Hepburn Springs, demonstrating his skill as a bush carpenter both with temporary buildings and scaffolding. While plans were being drawn, Harry began quarrying for the cellar, then moved on to a nearby slope. Working to the sound of birdsong, he joked that the pardelote was always telling him to 'give it up'. Harry, often with his pipe in his mouth, worked to an increasingly high standard as the walls went higher. He said the quality of the wall is most important at eye level. He would take some large flat stones and build them hanging out from the wall face, being sure to build them deep into the wall to balance the cantilever (Figure 2.2). Harry was very selective about which stones went onto the walls of the house, many of his rejects were built into garden walls, mostly by my mother and my sister Anne. After construction of the arch across the fireplace in the Hepburn Springs house (Figure 9.7), Harry moved with Pat to build again near Moruya in New South Wales.

Figures 4.2, 4.10, 4.11, 4.14, 5.4, 6.4, 11.6 – Doug Swing came from Colorado, USA, to Australia, where we met and got married. Doug had worked as a hard-rock miner and a carpenter. He took a keen interest in stone building, becoming a local expert at quarrying with hand tools. Together we developed our knowledge of stonework through visiting numerous stone houses in Victoria and New South Wales, talking with stonemasons and owner builders. Doug played a major role in supplying stone for the construction of the house near Hepburn Springs, and used his carpentry skills to make centre formwork for arches (Figure 7.28). After my mother moved away, he and I continued on the project, working on stonework for a round window (cover photo and Figure 7.26), and completing the walls.

Also pictured: Figure 2.9 – Doug with daughter Phoebe; Figure 11.6 – Doug on a building site with Bert Behnke, stonemason and builder, at Penshurst. We visited Penshurst when the drought broke in the 1980s, and he was helpful, telling us how he made chisels to suit his own needs. He showed us stonework done by his family company, with a feature of occasional stones sitting forward from an otherwise straight wall.

Figure 4.8 – pictured with Harry, Tom Manning, a neighbour in the Hepburn Springs area, answered our call for help when a particularly large stone was needed for a mantelpiece (Figure 9.8). Tom has done stonework on various other jobs (Figure 6.9). He also worked on joinery in the round kitchen for me, including cutting the benchtop to a curve by chainsaw (Figure 1.14), helped by another local, Steve Henderson.

Figure 7.2 – a relative of mine in Tasmania, John Gunn is pictured in the doorway of a stone folly built by his father, Fred Gunn. The Gunn family building company was able to re-use demolished materials, such as the lining for the window interior (Figure 7.3).

Figure A2.2 – a Castlemaine local, Seth, photographed working on a stone retaining wall.

Construction notes on photographs

Figures 1.1, 9.11, 9.12, 10.8, 10.9 – Camelot, 1938, architect Eric Nicholls, who also worked with Walter Burley Griffin and Marion Mahoney, carrying on the practice after the death of Burley Griffin.

Figures 1.15, 1.16, 5.14 – Johnson house, c. 1922, architect Walter Burley Griffin. Note that Marion Mahoney, his wife, worked closely with him and, while her drawings are well known, it is difficult to ascertain how much of the design work was also hers.

Figure 5.5 – Grant house, c. 1922, architect Walter Burley Griffin. Other work photographed in Castlecrag, New South Wales, where not described, is from the same architectural practice, either Griffin or Nicholls.

Figures 1.6, 1.7, 1.21 – Baker courtyard house, the first of two by architect Robin Boyd.

Figures 1.3, 1.4, 3.1, 3.14, 3.15, 11.3 – the second house for the same owner, by Robin Boyd, uses a paler stone, quarried on the property.

Figures 1.14, 1.13, 3.10, 4.1, 4.2, 4.4, 4.5, 4.7, 4.10, 4.11, 4.13 to 4.15, 5.4, 7.3, 7.6, 7.26, 7.28, 8.1 to 8.3, 8.17, 9.7 to 9.9, 9.15, 10.3, 11.5, A2.1 – house near Hepburn Springs, designed by me for my mother, Pat Gunn. Stonework was mostly by Harry Vanderburg. Later work was done by Doug Swing and me, including bull's-eye window (Figure 7.26) using pale stone recycled from a chapel tower restoration. Quarrying of stone on site mostly by Doug Swing. Balustrade of bush timber (Figure 11.5) made by Steve Henderson.

Figures 2.2, 4.8, 4.9, 4.16, 8.16, 8.18, 8.21 – stonework and granite quarrying for house near Moruya by Harry Vanderburg.

Figures 1.9, 1.23 to 1.27, 2.1, 7.15, 8.10 – Narrapumelap, in Wickliffe, Victoria, built for John Dixon Wyselaskie by architects Davidson and Henderson. The tower originally had a high roof. Wyselaskie farmed sheep in Wickliffe from 1840. In 1873 he had 50 stonemasons working for him. Between 1874 and 1876 he travelled to Europe with his wife, then in 1878 he moved to Melbourne because of poor health. The property was sold after his death in 1883 to G.N. Buckley, and in 1921 the land was subdivided for soldier settlement. The homestead was in disrepair but is now mostly restored.

Figures 2.13, 2.14, 3.13, 11.2 – Taradale stone laid by Les Pitt and Peter Matthews.

Figures 2.17, 2.18 – Narrandera stone buildings built by Jock McLeod.

Figure 6.9 – stonework by Tom Manning, for architect David Moore.

Figures 7.1, 7.20 to 7.22, 7.29, 7.33, 7.34, 11.1 – Abercrombie House, built for James Stewart, on a site where his father had built on an 1827 land grant. Architect George Allen Mansfield worked from a design in the book *Villas and Cottages*, Calvert Vaux, 1857 (an unbuilt proposal for Matthew Vassar by Vaux and AJ Downing).

Figure 8.19 – *Black Spring* sculpture by Andy Goldsworthy, 1992, constructed from Parachilna slate by Joe Smith.

Figures 2.18, 5.1, 5.11, 9.14 – many of the early sandstone houses in Hunters Hill were built on subdivisions started in the 1850s, by Charles Edward Jeanneret or the brothers Didier and Jules Joubert.

Figures 3.9, 6.4, 6.14, 7.11, 9.1 – a number of stone buildings in the Bairnsdale district are attributed to George Rawlings. The initials G.R. appear on a carved dedication stone at one, a farmhouse.

Owner built houses

Figures 1.17, 1.18, by Bob McIlroy.
Figure 1.20, by Norm Foster.
Figures 2.5, 8.23, 10.1, 10.4, by Sandra White.
Figure 3.7, by Brad and Jenny Pickering.
Figure 6.8, by Chris Taun.
Figure 6.21, by Bob Speed. The house was owner built, 'so I don't have to go off somewhere else to earn the money'.

Index

www.ingramcontent.com/pod-product-compliance
Ingram Content Group UK Ltd.
Pitfield, Milton Keynes, MK11 3LW, UK
UKHW051956041026
15590UKWH00009B/2337

9 780643 096370